MW00851237

Horticulture Management & Supervision

Horticulture Management & Supervision

Donald W. Jackson

Lighty Chair in Horticulture
and
Director and Instructor
Clarence W. Schrenk Program
in Horticulture, Landscaping,
and Turf Management

The Williamson School

DELMAR
CENGAGE Learning™

Australia Canada Mexico Singapore Spain United Kingdom United States

Horticulture Management and Supervision
Donald Jackson

Vice President, Career and Professional Editorial: Dave Garza

Director of Learning Solutions: Matthew Kane

Acquisitions Editor: David Rosenbaum

Managing Editor: Marah Bellegarde

Product Manager: Christina Gifford

Editorial Assistant: Scott Royael

Vice President, Career and Professional Marketing: Jennifer McAvey

Marketing Director: Deborah Yarnell

Marketing Coordinator: Jonathan Sheehan

Production Director: Carolyn Miller

Production Manager: Andrew Crouth

Content Project Manager: Kathryn B. Kucharek

Art Director: Dave Arsenault

Technology Project Manager: Mary Colleen Liburdi

Production Technology Analyst: Tom Stover

Library of Congress Control Number: 2007940811
ISBN-13: 978-1-4180-3998-1
ISBN-10: 1-4180-3998-5

Delmar Cengage Learning
5 Maxwell Drive
Clifton Park, NY 12065-2919
USA

Cengage Learning products are represented in Canada by Nelson Education, Ltd.

For your lifelong learning solutions, visit **delmar.cengage.com**
Visit our corporate website at **cengage.com.**

Printed in the United States of America

1 2 3 4 5 6 7 12 11 10 09 08

This book is dedicated to the women in my life: Mary Beth, Meredith, Anne, Joan, *and* Bertha.

CONTENTS

Preface ix

Acknowledgments xi

About the Author xiii

Chapter 1
Starting Out on the Right Foot: Being Successful in Your First Full-time Job **1**

Chapter 2
Time Management: Key to Success **11**

Chapter 3
Motivation: Essential to Surviving in the Green Industry **25**

Chapter 4
Decision Making: Making or Breaking Your Future with the Company **37**

Chapter 5
Communication: Relating to Those Around You **47**

Chapter 6
At the Head of the Pack: How to Be a Leader **58**

Chapter 7
Delegating Authority: How to Get the Job Done **71**

Chapter 8
Understanding the Big Picture: Managing the Fiscal Health of Your Business **84**

Chapter 9
Effective Meetings: Making the Most of Your Time **97**

Chapter 10
Planning: How to Be Most Effective in Your Position **108**

Chapter 11
Dressing for Success: Looking Your Part **118**

Chapter 12
Working Through Change: Developing as a Young Supervisor **127**

Chapter 13
Hiring Staff: Making the Right Choices **136**

Chapter 14
Evaluating Staff Performance: Overseeing Others in the Workplace **146**

Chapter 15
Forging a Career Path: How to Succeed Long-Term **156**

Bibliography 167

Index 169

PREFACE

Horticultural skills are critical to a successful future in the Green Industry. Moving up through the managerial ranks of an organization, however, requires fine-tuning supervisory skills. From delegating staff to acquiring leadership skills and coordinating day-to-day tasks with an organization's long-term goals, young supervisors must develop many talents.

Recent graduates must be proficient in their communication skills and adept at making timely and well-thought-through decisions. They should be practiced in managing and prioritizing their responsibilities and be able to plan for today, tomorrow, and well into the future. Young supervisors must likewise gain a good understanding of budgetary concepts and the financial information presented in income statements, cash flow statements, and balance sheets. They need to cultivate an understanding of the many intricacies associated with running a horticultural business and be attuned to the seasonality of the trade.

To remain competitive in today's workplace requires a breadth of managerial skills. For example, customer-service skills are closely linked to the long-term health and productivity of a business. Planning and overseeing a meeting or dressing professionally likewise require experience and expertise. Perhaps most importantly, shaping a future career in the horticulture profession and meshing long-term aspirations within the overall goals of the institution are not easy tasks.

The Green Industry comprises a broad assemblage of distinctly different trades and professions. To its credit, such diversity allows recent graduates a wide and exciting array of employment possibilities from which to choose. From the landscape, nursery, and arboriculture professions to the golf course and sports-turf industries, career options are virtually limitless. The greenhouse, florist, arboretum and botanical garden aspects of the trade offer further opportunities, and horticulture roles within zoological gardens, theme parks, and private estates continue to expand.

By discussing supervisory concepts in this textbook, I hope to make your future within the Green Industry that much more successful. Thank you for the opportunity of sharing these important concepts with you. I wish each and every student the very best in following his or her dream within the horticulture profession.

Donald W. Jackson
Lighty Chair in Horticulture and Director
and Instructor of the
Clarence W. Schrenk Program in Horticulture,
Landscaping, and Turf Management
The Williamson School

Acknowledgments

It is with sincere appreciation that I thank Mr. A. David Sharbaugh, owner of Sandia Technical Services, and my Williamson faculty colleague Mr. Dennis Johnson for reviewing the financial concepts presented in Chapter 8.

Mr. Joseph Hoplamazian Jr., president of Wedgewood Gardens; Mr. Thomas Speakman, president of Speakman Nurseries; Mr. Mark Cole, vice president of production at the Conard-Pyle Company; and Mr. George T. Smith, owner of Smith's Tree, generously shared their professional expertise. I would like to thank Mr. Hoplamazian, Mr. Speakman, and Mr. Cole also for graciously allowing me to photograph aspects of their operations.

I am most appreciative to Mr. Brian Frantum for legal assistance and to Ms. Anne Frantum, fellow faculty member at Williamson, for guiding me through manuscript stylistic and formatting nuances. Thanks go to Dr. Clifford A. McKay Jr., Mr. Clifford A. McKay III, and Mrs. Mary Beth Jackson for critical review of the manuscript, and I am indebted to Mrs. Bertha H. Jones for many hours of dedicated editorial assistance.

I am particularly thankful for the dedicated support of Mr. Wayne Watson and Dr. Paul Reid, chairperson of the board of trustees and president, respectively, of Williamson, and academic dean Mr. Thomas Wisneski. I am also appreciative of the continued support provided by Williamson trustee Dr. Richard Lighty, as well as the many past and present Williamson faculty and horticulture students who encouraged the writing of this book.

It is with sincere gratitude that I acknowledge the direction and leadership of Mr. David Rosenbaum and Ms. Christina Gifford, acquisitions editor and project manager, respectively, of Delmar Cengage Learning. Both were instrumental in encouraging and promoting this book to become a reality. And finally I would like to thank Kathryn B. Kucharek, Senior Content Project Manager, Delmar Cengage Learning; Maryann Short, the copy editor; and Arunesh Shukla, project manager, Newgen Imaging Systems, for their assistance throughout the development of the book.

Thank you.

Donald W. Jackson
Lighty Chair in Horticulture
and Director and Instructor of the
Clarence W. Schrenk Program in Horticulture,
Landscaping, and Turf Management
The Williamson School

The author and Delmar wish to thank the following reviewers for their time and content expertise:

Lynn DuPuis
Alfred State University
Alfred, NY

Jeff Iles
Iowa State University
Ames, IA

Gerald B. White
Cornell University
Ithaca, NY

Kim Marie Yates
Mayland Community College
Spruce Pine, NC

Warren Lytle
Shasta College
Redding, CA

About the Author

Donald W. Jackson has worked in the horticulture industry for over 25 years. He holds horticulture degrees from Finger Lakes Community College in Canandaigua, New York, and the University of Kentucky in Lexington.

Mr. Jackson has over 12 years' professional experience in public horticulture at the Cincinnati Zoo and Botanical Gardens, Zoo Atlanta, and the Kansas City Zoological Gardens. Throughout his tenure at these institutions, Mr. Jackson served in many capacities, including vice president and general manager at Zoo Atlanta. He has likewise directed horticulture, volunteerism, customer-service, and marketing programs.

Mr. Jackson has authored over 95 technical papers and articles in a wide array of professional and popular publications. He has published in Harvard University's *Arnoldia*, the London Zoo's *International Zoo Yearbook*, *Journal of Volunteer Administration*, *American Nurseryman*, *Nursery Management & Production*, *American Horticulturist*, *Early American Life Gardens*, *Saturday Evening Post*, and Delta Airline's *Sky*. He also authored a chapter in *Wild Animals in Captivity*, a book sponsored by the Smithsonian Institute–National Zoological Park.

For the past 13 years, Mr. Jackson has directed and instructed in the Clarence W. Schrenk Program in Horticulture, Landscaping, and Turf Management at the Williamson School in Media, Pennsylvania. He has likewise served on advisory committees at Longwood Gardens in Kennett Square, Pennsylvania, and the Tyler Arboretum in Media, Pennsylvania.

Starting Out on the Right Foot: Being Successful in Your First Full-time Job

OBJECTIVES

After studying this chapter, you should have an understanding of:

- developing strong professional relationships
- problem-solving capabilities
- support the long-term goals
- motivate full-time and seasonal staff
- interpersonal issues
- teambuilding
- leadership skills
- technical or managerial support

Although the majority of horticulture graduates are successful in their first full-time positions, some fail. Most of these failures have little to do with the recent graduate's grasp of horticultural principles and everything to do with how the individual fulfills his or her new role as a full-time supervisor.

More likely than not, supervisors' deficiencies have to do with how well they relate to coworkers, how well they motivate their staff, their ability to be punctual for work, their written and oral communication skills, their ability to delegate, and whether they meet deadlines. Performance failures might also be traced to supervisors' inability to plan and organize each day's responsibilities both for themselves and their staff.

In this chapter we discuss management techniques that new graduates need to learn in order to succeed in their first full-time position. The sooner you grasp these concepts, the quicker your rise through the ranks of any horticultural organization. Put into practice the information presented in this and ensuing chapters, and you will have a much greater chance for a successful professional career. ❧

Achieving success in your first full-time position often means

1. being persistent in seeing a goal through to its completion.
2. being proactive in completing responsibilities.
3. motivating yourself to reach your goals.
4. focusing on the long-term goals of the organization.
5. keeping an open mind and learning from your mistakes.
6. persisting in seeking answers to difficult questions.
7. not fearing failure, but being afraid of not trying.
8. learning to problem solve.
9. creatively seeking answers to difficult questions.
10. maintaining professional ethics and values.

Staff Management and Problem Solving

One concept which is crucial to the success of any young supervisor deals with the necessity of developing strong professional relationships with his or her staff. Although this is important for many reasons, it is especially critical when striving to fulfill the long-term goals of the organization. Young supervisors must likewise learn to empower their staffs toward meeting a well-planned and coordinated set of objectives. In order to be successful, managers must be able to rely on their staffs in order to follow through on key responsibilities. A young supervisor's ability to motivate his or her staff will critically influence the level of achievement throughout his or her professional career.

It is critically important for supervisors at all levels to be professional and proactive members of the management team. Although upper management does not expect recent graduates to have all of the answers, they expect them, as supervisory staff, to have problem-solving capabilities. Practice solving the problems that face upper-level managers to gain problem-solving skills. At the same time, you will have a better understanding of why some decisions have been made and how they support the long-term goals of the institution. What managerial concerns do you see as being of critical importance today, tomorrow, and well into the future? How would you cut financial waste and increase profit margins? How could short-term plans better mesh with the long-term goals of the organization? How would you provide better customer service? What could be done to more effectively motivate full-time and seasonal staff? How would you encourage staff to take more ownership of critical issues in their realm of responsibility and more responsibility for solving them? How could staff be more timely in completing critical tasks without sacrificing quality? What steps would you take to encourage management and staff to work more cooperatively together? How would you achieve better communication between supervisors and staff? What could be done to address each concern and how would the steps taken affect related areas in the organization? By exploring these and other crucial issues, young supervisors will hone their problem-solving abilities and increase their value to the organizational team.

Rather than reinventing the wheel when problem solving, taking an existing process or technique and modifying it to be more cost or labor efficient often has more value. New employees who are perceptive and closely observe their work-related environment might see a solution to an old problem. Often, successful problem solving involves little more than looking for answers that support the short- and long-term goals of the organization.

Another facet of problem solving is to understand the importance of coordinating with your immediate supervisor. Young managers are expected to go to their supervisor for guidance on issues under consideration, and they should also take potential solutions for those issues. Problems are much easier to solve if they have not been allowed to grow into a crisis.

Effective problem solving involves

1. being disciplined in one's risk-taking.
2. being able to clearly define the company's vision of its future.
3. strategically aligning oneself with organizational goals.
4. working successfully with all levels of management.
5. being financially responsible and staying within departmental budgets.

Goal Setting as Related to Problem Solving

It is impossible to solve a major crisis or concern without taking the short- and long-term goals of the organization into consideration. The setting of personal short-term and long-term goals helps young managers establish a road map to reach career aspirations. Goal setting's importance lies in helping us focus on our primary objectives and staying motivated. Hastily thought-through goals are little better than having no goals at all. Record your work-related goals and refer to and reevaluate them over time. Circumstances change, and the goals of importance today may or may not remain a top priority tomorrow. One's goals should likewise have the capacity of being accurately measured or quantified. This is particularly valuable in that it allows supervisors to track their progress in achieving their highest priorities. Professional goals should remain malleable so they can be amended over time. Flexibility in goals allows you to grow and develop in your managerial responsibilities as situations change. Finally, it is imperative that one's work-related goals be accurately recorded so that they can be referred to and re-evaluated over time. Circumstances change, and the goals which are of crucial importance today may or may not remain a top-priority tomorrow. Always have a specific time frame for completion of each goal. Goals should be quantifiable, allowing supervisors to track progress toward meeting them.

Being Part of the Team

Success-minded young supervisors should always work toward being the most motivated and diligent member of the team. Be fully accountable for your responsibilities and be supportive of other supervisors throughout the organization. Such a mind-set goes a long way toward encouraging other coworkers to work as a team, but keep in mind that all supervisors meet resistance from their staff from time to time. There will even be times when working through the interpersonal dynamics of an organization will seem like the hardest part of your job. Although such personnel-related issues may be arduous to sort out, the following suggestions should help everyone work together as a cohesive whole. Develop an understanding of how each staff member relates to others as a professional. Some avoid interpersonal issues, and others are seldom happy unless they are embroiled in some petty controversy. Avoid these latter if possible and work with those who are most dedicated to achieving the best outcome for the organization.

Teamwork is one of the most vital components to the financial health and overall productivity of any business. Teamwork results from teambuilding. It allows staff to corroborate in decision-making processes and is instrumental in supporting organizational change. Teamwork enables businesses to realize their short- and long-term goals. Moreover, it promotes a supportive relationship between management and staff. Employees who work as a team are essential to the future health and productivity of any company.

Encouraging a team-based philosophy is in keeping with the objectives of any forward-looking organization. This is especially true in seasonally dependent professions such as the landscape, nursery, turf management, or arboriculture industries because much must be done in a short time. Such organizations should establish high expectations for their staff and expect managers to conduct themselves as professionally as possible. Young supervisors must be especially strong in their commitment to work together for the good of the organization. Those who truly excel in these areas of responsibility are almost assured a quick climb up the managerial ladder.

Young supervisors should remember that

1. problem solving directly correlates with long-term success in the trade.
2. not becoming ensnared in petty controversies is a characteristic of respected managers.
3. a problem usually has more than one viable solution.
4. have a contingency plan in case your initial solution fails.
5. effective management means being firm yet flexible in decision making.

Managing Staff Effectively

Teamwork is born of strong professional relationships between supervisors and staff. Managers must be able to rely on their staff to follow through on key responsibilities and empower them to meet objectives. The ability to motivate staff influences managers' level of achievement throughout their professional careers.

Managers at all levels should seek to achieve above-average leadership skills. Good managers motivate staff to work together and perform to the best of their ability. They listen objectively to the concerns of staff, which encourages staff to cohere as a team and support organizational goals. Consistently setting high and quantifiable standards for staff is a prime component of sound supervision. Praising them when they meet those standards and for supporting organizational goals is an often-overlooked component of becoming a first-rate supervisor.

Mentoring

A mentoring program encourages young supervisors to work more effectively within the management structure of an organization. To maximize the benefits of mentoring, ask the proper individuals for guidance. Don't request support from undependable coworkers or others who are overwhelmed with their own professional responsibilities. Carefully define your needs, especially if specific technical or managerial support is required. Make a specific request if you seek guidance on a specific issue. Incomplete or unclear requests delay solutions, and generic, blanket requests for assistance are difficult to answer. Before asking for help, do your part by analyzing the problem and listing the solutions you find. If that doesn't solve the problem, then request help. Briefly explain why you are soliciting help and tell the coworker your time frame for solving the problem. Finally, thank your coworker for the guidance and inquire how to return the favor in the future. The simple act of thanking coworkers will go a long way toward building quality relationships and proving yourself to be a mature and competent professional.

Working with Your Supervisor

One of the best ways to make yourself a valuable member of the organizational team is to anticipate the expectations of your direct supervisor and assess how you can best meet his or her needs. The easier you make your boss's job, the more integral your role in the company will likely become. Although it is perfectly acceptable to go to your supervisor with difficult issues, don't go empty-handed. Always have a complement of potential solutions. Make a habit of foreseeing problems early on and preparing possible solutions. This is an important concept and should be embraced by supervisors at all managerial levels.

A first-rate supervisor

1. follows through on responsibilities.
2. reliably supports other members of the team.
3. does not make excuses for failure.
4. understands that not making any decision at all is often worse than making a bad decision.

Another way to nurture a positive working relationship with your direct supervisor is to maintain confidence in yourself. By keeping your professional aspirations closely aligned with the long-term goals of the organization, you can work to the very highest levels of your ability to meet them. Taking full responsibility for your actions and consistently maintaining a mature and positive attitude will greatly increase your managerial value to any organization.

What if, as a new hire, you find you have a truly difficult immediate supervisor? Maybe your supervisor has occasional outbursts of rage, makes sarcastic comments about subordinates' work, or has wild mood swings. To best deal with difficult personalities until you can move on to the next level of your career, it is important, first, to understand that the wrath, erratic behavior, deception, or other conduct should seldom be taken personally. Always treating them with respect, even when you feel they don't deserve it, and acting in a professional manner may help avoid setting these people off. If negotiating with the person has failed to improve things, try to anticipate and plan for this person's behavior. Be aware that any loyalty you offer a deceptive boss will not be appreciated or rewarded. Be a realist and acknowledge that many deceptive bosses can be quite covert. Other members of the organization may have little clue as to how staff is being treated. It is not uncommon in such situations to be unable to document or quantify the behavior. Although such supervisors can make your work environment difficult, try to maintain a positive outlook and encourage change.

Managing staff to work as a cohesive and productive unit includes

1. encouraging them to take ownership of their job-related responsibilities.
2. allowing them to have a voice in the orchestration of assigned projects.
3. supporting their technical and professional needs.
4. sincerely telling them that they contribute to the overall decision-making process.
5. providing an environment in which they can problem solve as effectively as possible.
6. promoting their pride in significant accomplishments.
7. adhering to measurable and quantitative goals.
8. aiding them in their understanding and appreciation of the long-term goals of the organization.

Continued Education

Further education may involve enrolling in an accounting or personnel management course at a local college. It may mean pursuing a master's degree in business administration. Rising on the organizational ladder is more likely when training and education closely relate to the actual management of the business. Education is an important element in long-term success and in staying competitive in today's ever-changing marketplace.

Although learning can and often does take place in a collegiate setting, it does not have to. Learning continues long after attainment of a university degree, and knowledge gained after college will figure largely in your future advancement. Those who never participate in a professionally oriented class sponsored by their state's Cooperative Extension Service, do not stay abreast of advancements in the trade, and never attend horticulture or business-related workshops, seminars, or conferences almost assuredly will languish on a low rung of the organizational ladder.

Business owners know they need to stay current on all the new insecticides, herbicides, and fungicides that enter the marketplace. They likewise need to keep pace with Occupational Safety and Health Administration (OSHA) requirements. From worker compensation issues and immigration responsibilities to health and liability insurance for employees, supervisors at all levels need to be up to date in their certifications and licensing responsibilities. Managers in the Green Industry must know and abide by local, state, and federal laws.

Staying Competitive

Here are three final considerations for strategically managing your career and maintaining a competitive edge. First, do not allow your career to stagnate, but plan your career moves. The savvy supervisor understands that some opportunities are well worth investigating; others are best left to someone else. Second, grow and learn from each opportunity. Constantly stretch yourself to perform at the very highest level and strive for excellence in everything that you do. Finally, never compromise your integrity. Following the masses is not always the best path. A willingness to forge a new path and create the unexpected will serve you well.

ONCLUSION

The Green Industry can be a most rewarding and satisfying career. It is an extremely broad and ever-changing profession. The horticulture field challenges even the brightest and most inquiring minds, and all students need to be as well prepared as possible.

The remaining chapters in this book take a much more in-depth look at a wide array of managerial topics. Understanding them can better prepare you to be a supervisor in the Green Industry.

DISCUSSION QUESTIONS

1. What can supervisors who can think on their feet and anticipate problems offer to an organization?
2. How can young supervisors learn to think like an upper-level manager within an organization?

   Why is this an important goal?
3. Managers should expect staff to take ownership of company-related problems and help develop solutions to them. Why is this expectation inextricably linked to the future growth and development of an organization?
4. Why should new managers strive to anticipate company needs before they go to their supervisors with potential solutions?
5. Why does a young supervisor's promotability within an organization hinge on his or her ability to solve problems before they grow into major catastrophes?
6. If your supervisor believes that he or she has to direct your every move 24/7, what does this imply about his or her confidence in your managerial capabilities?
7. How does working as a team contribute to the health and well-being of an organization in today's Green Industry?
8. Why will understanding and working through the interpersonal dynamics of coworkers be the hardest part of your job at times?
9. Why should long-term goals and aspirations be malleable?
10. Why should a supervisor's professional aspirations be aligned with the long-term goals of the organization?

John Parker's Employment With Ricardo's Landscape Service

John Parker was hired as a full-time employee with Ricardo's Landscape Service, an installation and design-build firm based near Williamsburg,

Virginia. John had graduated this past spring with a degree in horticulture from a major midwestern university. He had worked with Ricardo's during the summer between his junior and senior years to fulfill the requirements of the department's mandatory internship program. He had graduated with a 2.54 overall-grade-point average (GPA) and received a "reasonably favorable" evaluation from Ricardo's upon completing their summer internship program.

John was hired as a crew member on a landscape installation team. As long as John's work on the landscape crew was satisfactory, he had the potential of being promoted to the position of crew foreman following his 6-month probationary period and on-the-job training. If he were promoted, John's compensation package would be adjusted to reflect his new and expanded responsibilities with the company. The probationary period was clearly outlined in the company's *Employee Handbook*, which John received from the director of Human Resources when he was hired. The entire hiring procedure and related probationary period was also outlined orally to John by the firm's director of Human Resources.

John was not a particularly enthusiastic or well-liked member of the crew; he required prompting before he would take on tasks and he did not get along with some team members. His performance did not improve appreciably over time. Some of John's interpersonal and productivity problems could be traced to his brash and know-it-all attitude.

The crew foreman sat down twice with John in his probationary period to review his job performance. The second time, they discussed John's performance as documented by the foreman. Although the written evaluation noted a number of John's attributes, it was by no means wholly positive on his overall performance. Although Ms. Ramirez did not have day-to-day contact with John, he had observed many of the deficiencies noted by the crew foreman and totally endorsed the written summary of John's performance. The evaluation praised John's punctuality, honesty, and level of horticultural knowledge. The review noted that he was deficient in motivation, would not always cooperate with other members of the crew, and had occasional outbursts of inappropriate language. He seldom accepted constructive criticism from his crew foreman. Irina Ramirez, a 17-year veteran with Ricardo's and its director of Landscape Services, was also present at John's review. Although Ms. Ramirez did not have day-to-day contact with John, she had observed many of the deficiencies noted by the crew foreman and stood behind the written summary.

A few weeks after John's review, he was caught by the crew foreman and the director of Landscape Services with a beer while on the job. John was drinking from a bottle concealed in a brown paper bag while sitting in the cab of the company's pickup truck. This was a clear violation of a chief prohibition in the Employee's Code of Conduct outlined in Ricardo's *Employee Handbook*. John's misdeed was reported to the director of Human Resources and his employment was immediately terminated.

QUESTIONS FOR DISCUSSION

1. What were the three major issues in John's failure to perform as a competent and responsible employee with Ricardo's Landscape Service?
2. What should John have done to improve his deficiencies as a full-time employee with the company?
3. Which of John's personality traits was most damaging to his position as a landscape crew member with Ricardo's?
4. How significant was John's drinking as a deciding factor in his future with the company?
5. How would you have reacted as crew foreman and direct supervisor of John in this situation?
6. How would you have reacted as director of Landscape Services in this situation?
7. How would you have reacted as a motivated and career-oriented crew member who had to work closely with John each day? Think through this question carefully and give a detailed description.
8. Did Ricardo's Landscape Service use good judgment in hiring John as a full-time employee in the first place?
9. How valuable was John as a summer intern with the company?

CRITIQUE

Expectations of companies paying top wages to new horticulture graduates are high, and expectations only increase as the workplace becomes more and more competitive. Recent graduates who fail to meet these expectations move up through the ranks slowly if at all. They also may not receive the year-end raise they hoped for. Unlike John, new employees usually last more than a few short months with their first employer. A situation such as that described here, an employee fired for misconduct during his probationary period, is a more extreme example of failed expectations, but it is certainly worthy of consideration. Let's fully dissect John's employment scenario and see if we can arrive at some solid conclusions as to why he failed.

John displayed no problem-solving abilities and did little to think like an upper-level manager in his new position with the company. John's overall attitude was much of the problem. John lacked in his ability to be a team player, made little effort to coordinate with his direct supervisor, and would not accept constructive criticism—a very basic part of learning on any job. Most importantly, John showed no interest in improving. He had had ample oral warnings and written documentation from his crew foreman with regard to his on-the-job deficiencies. Even so, he continued swearing, a relatively simple behavior to reform but a prime factor in customer dissatisfaction, in residential and commercial landscape settings. That Ms. Ramirez, the director of Landscape Services, participated in John's performance review should have warned John that his job was in jeopardy.

John's positive attributes of punctuality, honesty, and horticultural knowledge were not enough to compensate for his deficiencies. The drinking on the job simply expedited his exit; he was on his way to termination anyway. John took no initiative to meet with his supervisor and amend his ways by mapping out a plan for improvement. If John had developed goals associated with his long-term future with the organization, he might have avoided being terminated. Finally, John made virtually no effort to take responsibility for his actions. It is one thing to make mistakes, it is quite another not to learn from them.

Managers at Ricardo's Landscape Service were professional in their treatment of John and gave him several opportunities to improve his performance. He received oral and written warnings, and their director of Human Resources had provided John with the *Employee Handbook* and other tools needed to be successful with the company. It was up to John to take advantage of the help the company offered. However, the company could have saved itself a headache by evaluating John's internship more carefully before offering him full-time employment. His final internship evaluation had been far less than stellar and should have suggested many of John's behavioral shortcomings as a full time employee. As we shall discuss in later chapters, a student's internship experience should be an intense learning experience. An internship that meshes with the student's long-term professional goals has an excellent chance of benefiting employer and employee, often leading to full-time jobs.

Time Management: Key to Success

OBJECTIVES

After studying this chapter, you should have an understanding of:

- seasonality
- organizing and executing around priorities
- wholesale B&B nursery
- gross annual income
- purchases on account
- short-term responsibilities
- long-term goals
- freedom of expression and creativity

Go to any well-stocked bookstore, and you will find shelf after shelf of management and motivational texts that deal with time management. The seasonality of the Green Industry makes time management a must-have skill for supervisors; every good manager plans for the busy season during slower times.

Northern landscape, nursery, turf management, and arboriculture companies must earn a significant percentage of their yearly income in a few short months. Greenhouse growers unable to capitalize on holiday sales would never survive. Poinsettias are not sold at Mother's Day and Easter lilies are not sold at Thanksgiving. Imagine the effect on yearly revenue of a retail florist who does not provide red long-stemmed roses to customers on Valentine's Day. The season with the greatest impact on the horticulture profession might well be the few short weeks that Christmas-tree growers have to cut their pines, spruces, and firs and send them to market. A garden center's year-end sales would be depressed without purchases of holiday trees from Thanksgiving through Christmas.

Back when large hand-hewn stones formed building foundations, accurately laying the cornerstone was the most important step. If the cornerstone were not correctly set, construction of the foundation and ultimately the overall integrity of the entire structure would be compromised. Developing a firm foundation of time-management skills is no less vital to your future. The extent to which a supervisor develops these skills will largely dictate the success of his or her career. ❧

For effective time management,

1. Keep track of your time.
   Just as weighing yourself is the first step in trying to lose weight, tracking how you spend your time is the first step to managing it.
2. Review how you spend your time.
   Reviewing and analyzing your daily schedule allows you to see how your time is used and to determine whether changes in focus are appropriate.
3. Be more efficient in how you spend your time.
   Analyze your time usage and make the changes that you think will improve your time management.
4. Constantly reevaluate your time-management skills.
   Reevaluate and adjust how you manage your time to best serve your priorities.

Prioritizing Tasks

One of the most important concepts of managing time is organizing and executing around priorities (Covey, 1990, p. 149). Many examples in the horticulture trade could be used to illustrate this principle, but the wholesale ball-and-burlap (B&B) industry does it best. It is not uncommon for a wholesale B&B nursery to earn as much as 60% of gross annual income in as little as 8–10 weeks each spring. Although B&B nurseries can often count on a second surge of business in the fall, the large majority would find it difficult if not impossible to salvage their fiscal year if spring revenues fell significantly below expectations.

The obvious first priority for an owner of a wholesale B&B nursery is to organize and use time as effectively as possible during March, April, and May. An owner should try to maximize profits at this time of year as there is little margin for error should the spring digging season not go according to plan. To organize and execute around this priority and help assure that the spring digging season will proceed as planned, weeks if not months of preparation are needed. Skid-steer loaders and related digging equipment must be prepared for the busy months ahead. This means not only a complete servicing of the engine but also a thorough check of tires, hoses, and other components. The same goes for trucks, tractors, trailers, and other in-house equipment that directly or indirectly support the digging operation. Major equipment failures during the spring digging season result in a critical loss of revenue for the nursery. With planning, maintenance oversights are avoided.

Fill all staff positions—management, digging crews, and related labor positions—with experienced and well-qualified personnel. The nursery should be stocked with wire digging baskets of various sizes, squares of burlap, and boxes of sisal twine. The office staff must know to coordinate orders with those overseeing the digging operation. They should be adept at processing invoices and keeping abreast of overdue purchases on account of trees and shrubs. Additionally, the office staff should be able to courteously and professionally deal with a wide range of customer-service issues. To prepare himself or herself for the spring digging season at a wholesale B&B nursery, a young supervisor asks well-thought through questions. Seeking guidance is not only a fine way for the novice to learn but also assists the nursery owner in determining needs for this financially critical time of year.

Landscape companies and garden centers account for a large percentage of tree and shrub orders placed with a wholesale B&B nursery. Municipalities, universities, and similar organizations request plant material for on-site landscape installation projects. Some wholesale nurseries custom dig plant material for a very specialized and select group of customers, but most B&B operations are far more diversified and provide a wide range of trees and shrubs to their clients. If a wholesale B&B nursery does not have its own fleet of semi-tractor trailers and in-house drivers, it will need to coordinate its transportation needs with an outside trucking firm. Because most nurseries ship to clients many hundreds of miles away, it is imperative that reasonably priced and dependable transportation be coordinated well before the onset of the spring digging season.

The owner of a wholesale B&B nursery will need to orchestrate many additional responsibilities to organize and execute around priorities in preparing for the spring digging season and ensure a well-managed and successful year.

Four Core Principles

The four simple core values of effective time management listed here are essential to the day-to-day success of managers at all levels. They help supervisors manage for the short term, yet allow them to keep the big picture in sharp and constant focus. Supervisors who follow these principles are better able to handle many of life's smaller issues.

1. Time management must be well coordinated.
   Supervisors cannot efficiently handle managerial responsibilities if they do not follow time-management principles elsewhere in their lives.
2. Positive changes must take place in your professional life.
   Managing time must start with a sound evaluation of how a manager can make his or her professional life more productive. Young supervisors must be diligent in prioritizing their day-to-day responsibilities. They should also develop contingency plans for times when things don't go as planned.
3. Finding solutions to larger problems is an important component of time management.
   It is easy for a person to make an excuse when a task is not accomplished as planned, when often the real reason is that time was not efficiently managed. For example, perhaps more tasks could be delegated. More productive use of your managerial time will come with time-management analysis.
4. Time is finite.
   Remember that there are 24 hours in a day and 7 days in a week. Use your time as wisely as possible. Managers cannot regain even 1 second that may have been wasted. Instill this rule in your staff as well.

Planning

Time-management techniques cannot be divorced from other managerial concepts, and in fact are intimately related to delegation and planning. Planning allows supervisors to set long-range goals and organize and execute around priorities. To truly manage time, supervisors constantly plan (Figure 2-1).

Time-management skills often reflect how you live your life.

1. Taking control:
   You control how you spend time in your professional and personal life. The way you manage your day-to-day responsibilities while on the job reflects your overall values and commitment to the organization.
2. Reaching goals:
   Stretch yourself to reach goals truly worth achieving. Unmotivated individuals are seldom able to achieve their maximum potential, whether in their personal and professional lives.
3. Improving self-esteem:
   How much you accomplish in life hinges on your self-esteem. Those who fail to accomplish much often have low self-esteem. How effective are they in managing their time?

Continued

Continued

4. Being flexible: One of the most valuable traits to have when managing time is flexibility. Solving a problem tomorrow may take different strategies than those successfully used today or yesterday.

Activities for next week (August 23–27)

Top priority

1. Repair end zone on football field that was damaged by vandals—use sod
2. Weed flower beds near administration building (complaints received)
3. Routine mowing
4. Fertilize lawn areas and athletic fields
5. Aerate softball field
6. Seed area (1,200 sq. ft) dug up by telephone company
7. Maintain buildings with stringtrimmer
8. Cut down dead maple by library
9. Overseed lawn at president's residence
10. On Friday, do activities 1, 2, 3, 4 on rainy-day list if not already completed

Rainy-day activities

1. Take truck to Bill's garage for inspection (sticker expires Aug. 31)
2. Adjust carburetor on turf vehicle
3. Check sprayers (need them for weed control week of Aug. 30–Sept. 3rd)
4. Meet with employees for training session on pesticide safety and to discuss ways to improve appearance of campus (1 hour)
5. Pick up peat moss in Albany
6. Mower maintenance (sharpen blades, lubricate, change oil, replace belts on Thompson riding mower)
7. Update supply inventory
8. Paint employees' lunch room

Monday, August 23

Joe B.	8–12	Cut sod, repair football field end zone
	1–5	Fertilize athletic fields
Sally K.	8–10	Work with Joe on sod
	10–12	Mow athletic fields
	1–5	Continue mowing
Al M.	8–10	Weed flower beds by administration building
	10–12	Fertilize lawns
	1–5	Continue fertilizing
Heather S.	8–10	Work with Al on flower beds
	10–12	Mow lawn areas
	1–5	Continue mowing

note–1/2 of mowing done

Sod job looks good

All athletic fields are fertilized

[The daily work schedule may also be prepared without assigning approximate completion times, as shown in the following schedule.]

Tuesday, August 24

Joe B.	1. Irrigate athletic fields
	2. Seed area dug up by phone company
	3. Core-cultivate softball field
Sally K.	1. Seed area dug up by phone company (Joe will help)
	2. Mow
Al M.	1. Finish fertilizing
	2. Trim around buildings and fences with weedeater
Heather S.	Mow

FIGURE 2-1 Example of a work priority list prepared by a supervisor

You need to set aside ample time for planning. Some supervisors seldom make a conscious effort to plan for today, to say nothing of planning for tomorrow, next week, or even next month. Some almost boast that they are too busy to plan. As a result, they waste a good deal of the workday. Working hard is admirable, but it should never take priority over working efficiently.

To reinforce the relationship between time management and planning, let's consider again organizing and executing around priorities using the previous example of a wholesale B&B nursery. We have already established that the spring digging season is an extremely high priority at most wholesale B&B nurseries. Although digging trees and shrubs from March into May is critical to the cash flow and ultimate success of the business, the owner of a wholesale B&B nursery has additional responsibilities. For example, wholesale nurseries replenish their tree and shrub inventory by planting lining-out stock. Although some owners plant the majority of their lining-out stock in the fall, many wait until spring. Especially in more northern states, there are sound reasons for planting lining-out stock in the spring. Lining-out stock planted in the fall has precious little time for its roots to knit into the surrounding soil. As a result, frost heaving with the onset of colder weather can thrust a healthy percentage of the plants out of the ground. The exposed roots desiccate quickly and the plants die. By contrast, lining-out stock planted in spring has ample time for its roots to knit into soil and resist frost heaving later in the year. Although trees and shrubs should be dug and shipped in the spring, planting lining-out stock at this same time assures the long-term health of the wholesale nursery business. Spring digging season at a wholesale B&B nursery and planting of lining-out stock go hand in hand with organizing and executing around priorities. Both of these responsibilities involve a significant amount of coordination for the entire nursery operation to run efficiently throughout this most critical time of year.

But wait, there's more. If the owner decides to install lining-out stock during the spring digging season, that entails deciding months earlier which species and cultivars to select and how many of each to order. If flowering cherries, magnolias, and crabapples are currently popular with designers and landscape architects, will they remain salable 2, 3, or more years into the future? The astute nursery owner is not only attuned to current desirability of specific plants when ordering lining-out stock but also aware of how likely they are to retain popularity until scheduled for sale.

A wholesale B&B nursery owner should have an idea of whether he or she is better off with rows of flowering cherries, magnolias, and crabapples or whether their market will have softened by the time they are ready for sale. The owner might be better off ordering more silverbells, serviceberries, or redbuds. Quite obviously, this is by no means limited to small flowering trees but applies to red maples, lindens, pin oaks, viburnums, hydrangeas, pines, arborvitaes, and evergreen hollies as well. Stories abound in the nursery industry of species and cultivars that had to be bulldozed out of the field a few years after planting because their popularity had not just waned but disappeared.

Evaluate your time-management skills by answering the following questions:

1. Do you have priorities that tend to change constantly or are not well defined?
2. Are you continually running from one unfinished task to another?
3. Do you avoid setting deadlines?
4. Do you set out to accomplish more than is humanly possible and become frustrated at your lack of accomplishment?
5. Are you constantly duplicating your efforts and others'?
6. Do you get overwhelmed with small details and miss the big picture?
7. Are you unable to say no to requests from others?
8. Do you often wing it rather than follow a defined plan?

Continued

Continued

9. Are you indecisive or slow in making decisions?
10. Are you a perfectionist?

A further detail to consider is that a shade tree of 2.0 in.–2.5 in. caliper is a very satisfactory size for most wholesale B&B nurseries to sell. It is affordable for the average homeowner or commercial business and large enough that clients feel they have received a substantial piece of goods in return for their hard-earned cash. At the same time, its root-ball diameter of between 24 in. and 28 in. makes it reasonably easy to carry across a client's property using only a ball cart, and nursery owners get a reasonably fast cash flow.

Although a tree of 3.0 in.–3.5 in. caliper is still a reasonable size—a red maple or pin oak has a root-ball diameter between 32 in. and 38 in.—it doesn't take long for most shade trees to outgrow their usefulness in a nursery's field. For example, an 8.0 in.-caliper shade tree has a much larger and heavier root-ball, approximately 80 in. (Davidson, Peterson, & Mecklenburg, 1994, p. 159).

Now let's compare the weights of these three sizes of B&B shade trees. Whereas a maple or oak with a 2.0 in.–2.5 in. caliper can be carried in a ball cart across a client's property, a larger shade tree with a 32 in. root-ball weighs close to 500 pounds. A shade tree with a 38 in. root-ball weighs in the neighborhood of 1,000 pounds. Shade trees with an 80 in. root-ball exceed 8,000 pounds, clearly beyond a size easily transported across a client's property (Davidson et al., 1994, p. 476). A crane or large backhoe is often necessary to unload a shade tree with even a midsize root-ball. Larger trees are prohibitively expensive in themselves for the average homeowner or small business and frequently require further expenditures for transport to the planting site.

Goal setting as a function of time management allows managers to

1. determine the goal that is most important to them.
2. outline the methods for best reaching their goals.
3. quantify successes or failures throughout each step of attaining their goals.
4. budget for the financial resources needed to attain a goal.
5. assure that a reasonable time frame has been established for meeting their goals.

Continued

Setting Goals

Effective managers constantly oversee their short-term responsibilities while keeping a watchful eye toward the future and long-term goals. Long-term accomplishments are inextricably linked to how well you manage the seconds, minutes, and hours of today. A manager remiss in setting goals almost certainly wastes time. Goals are important for many reasons, not the least of which is that they allow us to allocate the use of our precious hours, weeks, and months with a predetermined focus. Purposefully setting goals enables better crisis management.

An equally important aspect of time management is to commit to attaining your goals. It doesn't matter how well you allocate your time if you have limited commitment to achieving a goal. The commitment needed by an Olympic swimmer or a cyclist in the Tour de France greatly exceeds that required of the average person attempting to lose 10 pounds in 6 months. Be certain that goals are worthy of the time and effort needed to reach them.

Part of planning your goals is to write them down and prioritize them. Assign a reasonable time frame to guide you and your staff toward a goal's completion. Remember, goals are useful only with an associated time frame. A work-related goal could be to increase sales by a specific percentage by the end of the year or earn a promotion to the next supervisory level in the next 6–9 months. A personal goal could be making first-team honors as a college all-star in lacrosse or working for a 3.5 GPA or higher in the current academic year. Finally, you need an unbiased way to measure progress while working toward your goals.

Continued

6. create a contingency plan.
7. continually reevaluate the importance of each goal.

Managing Time

How many times have you heard someone say, "Someday, I'm going to . . ."? We can all finish the sentence with a long list of examples. Most of us dream of what we will accomplish in our professional lives, but a dream does not materialize unless it is planned as a goal and has time budgeted toward its attainment. To work productively, we must constantly reevaluate how we spend our time. We will probably not have more time in our work schedule tomorrow, next week, or next month than we have today. This is a difficult but very necessary lesson to learn. Too few managers make a firm commitment to manage time.

Fortunately, many constraints on our time can be anticipated and controlled to a large degree. Remember that you and you alone are truly responsible for how you manage your time. Most of our time constraints can be divided into those directly controlled or influenced by others and those that we control. For example, office communication is a two-way street and therefore not under our total control. We can be extremely efficient in e-mails, memos, and reports only to be slowed by others who are less organized and motivated. By contrast, when we arrive at work in the morning and when we leave each evening are in our direct control. As supervisors, we face a broad array of time-wasting interruptions, from unexpected telephone calls and innumerable e-mails to visitors who drop by without appointments. To manage our time effectively, we have to look within for the most control. This often demands that young supervisors be ruthless in adhering to their primary goals and responsibilities. Excuses as to why the 30 *Viburnum nudum* 'Winterthur' or 50 *Fothergilla gardenii* 'Mt. Airy' were not shipped to the landscape company on time are not likely to be accepted during the spring digging season at a wholesale B&B nursery.

To-do lists as a part of time management should be

- targeted toward tasks critically related to your day-to-day management responsibilities.
- in keeping with the long-term goals of the organization.
- accurately prioritized.
- reasonable with respect to the time frame for their items' completion.
- updated regularly for relevance.
- constantly reevaluated.

Most supervisors would benefit by delegating more effectively. Doing everything yourself does not allow your staff to learn and more importantly robs you of valuable time for tasks that staff doesn't have the skill or experience to complete. The well-known merchandiser J. C. Penney, founder of the department store chain, said that "the wisest decision he ever made was to let go after realizing that he couldn't do it all by himself any longer. That decision, made long ago, enabled the development and growth of hundreds of stores and thousands of people" (Covey, 1999, p. 171).

Another significant impediment to time management is procrastination. Although young supervisors may not enjoy many of the responsibilities delegated to them, putting off tasks frequently makes them more difficult. First, the job may expand in magnitude with time and take even more time to complete later than it would take now. Second, a client, supervisor, or peer is likely to be annoyed by the delay. Most supervisors should not work as an island, isolated from others. A spider web illustrates how our professional decisions and actions connect with and impact others around us. Third, completing the delayed job now will take time from current projects, which may have looming deadlines.

Some believe that time management diminishes an individual's freedom of expression and creativity. Actually, it supports freedom of expression and creativity. Without time management, supervisors would forever be putting out brush fires and have little time to act or think creatively. Another misconception is that supervisors are more efficient and decisive when working under the pressure of a fast-approaching deadline. However, many who hold this philosophy are procrastinators and trying to cover up a deficiency in their time-management skills. No topflight manager would suggest that completing jobs at the 11th hour results in quality work. Others believe that managing their time effectively will take away from their other responsibilities. The owner of a wholesale B&B nursery will be pulled and tugged in many different directions during the spring digging season. He or she must learn to focus on the priorities and hone time-management skills to perfection. Supervisors who manage their time well are the ones who complete their responsibilities.

Do not become overburdened with trivial jobs that take more time than they are worth. The seasonality of much of the horticultural industry means that supervisors at all levels are pulled in many different directions during busy times. As the weather begins to break in the spring, it sometimes seems that all of the firm's customers want their projects completed as quickly as possible. To avoid overload, keep two principles in mind. First, always focus on the big picture and work toward your long-term goals. This will prevent you from getting sidetracked and losing sight of your highest priorities. Second, make sure that your tasks as well as those delegated to staff are top priorities. Supervisors should establish firm priorities for themselves and their crew members and stick to them. No manager can complete every task or solve every problem at the highest level of thoroughness.

Time management that affects the scheduling of other crews or equipment in a company sharing resources can have a direct negative effect on a horticultural business. For example, suppose you arrive an hour late for work the morning your crew has been allotted use of the firm's only dump truck until 11:00 a.m. at the Alawar residence. Your tardiness will delay a second crew scheduled to use the truck later in the day. The second crew will not be able to complete its landscape responsibilities at the Alawar residence by the end of the day and will have to return the following morning. Rosette's Catering Service, scheduled to cater Mrs. Alawar's poolside garden party the following day, will be delayed by the rescheduled planting of trees and shrubs adjacent to the pool. Mrs. Alawar, annoyed by having to delay her party, chooses another landscape firm to complete the second phase of her renovation. Your boss, knowing you were the ultimate source of the company's compromised reputation, fires you.

Organization

If your desk is cluttered or you can never seem to find the right file or invoice when needed, you need organization to avoid wasting time. Unorganized supervisors can be found in any organization, but not many will be found in middle- and upper-level management positions. Many managers use handheld organizers to help them manage their time. However, what you use to organize your time makes less difference than using whatever works well for you.

CONCLUSION

Although there are many ways to manage time, the core concepts are really quite simple. Track your time and review how you spend it throughout the workday. Take control of your time-management opportunities in your personal and professional life. Flexibility is as crucial to managing time as it is to reaching work-related goals. Never forget that time is a finite commodity, and be diligent in integrating planning into your repertoire of proven time-management techniques. Avoid wasting time and be more efficient in managing day-to-day responsibilities by not procrastinating, avoiding indecisiveness, and cultivating self-discipline. Do not be preoccupied with outside activities and avoid a perfectionism that interferes with goals. Finally, organize and execute around priorities.

DISCUSSION QUESTIONS

1. How does the seasonality of the Green Industry make time management more important for supervisors at all levels?
2. Do you believe that the adage "Time is Money" is a useful and viable concept in the landscape and nursery industries? Support your answer.
3. Which of the following managerial skills is most important? Why? How do these skills relate to each other?
   a. time management
   b. motivation
   c. delegation
   d. planning

4. What might some consequences be for a manager in the horticulture profession who lacks time-management skills?
5. Do you believe you have a significant degree of control over the time invested in your professional and career goals? What is the basis for your belief?
6. Why is it important to constantly reevaluate how to spend your time as processes and priorities change? Defend your answer.
7. Young supervisors need to delegate. Do you agree that it is important to work hard but not at the expense of working smart? Explain your answer.
8. Can you be an efficient and effective supervisor and be disorganized in your work habits?
9. Why do you think Dr. Covey so strongly advocates organizing and executing around priorities when managing your time?
10. How does finding solutions to larger issues often make the brush fires far less demanding?
11. Do you agree with the opinion of some supervisors that managing their time takes too much away from other job-related responsibilities? Explain.
12. Why must supervisors avoid being overwhelmed with the unending volume of information so readily available today? Explain.
13. Can a landslide of information slow a decision-making process and inhibit positive and proactive time management? Explain.

Anne Sawyer's Development as an Assistant Field Superintendent at Farmington Nurseries

In late summer, following her graduation with a degree in ornamental horticulture from a large midwestern university, Anne Sawyer accepted the position of assistant field superintendent at Farmington Nurseries in Kent County, Maryland. Farmington is a midsize nursery of approximately 600 acres, across the Chesapeake Bay from the Aberdeen Proving Ground on Maryland's Eastern Shore. Although Anne did not specialize in nursery management, Farmington Nurseries felt that her high collegiate GPA, sound letters of recommendation, and the enthusiasm and motivation she had exhibited in her job interview would make her a good fit for the job.

Anne was assigned to assist Rusty Sutter, a veteran of over 30 years in the nursery industry. Rusty had a 2-year junior college degree in nursery production and had been working in the nursery trade before his graduation from high school. For the last 17 years Rusty has been the field superintendent at Farmington and is highly respected as one of the premier growers in the tristate area. He is known throughout the industry as a fair but no-nonsense supervisor who expects diligence and attention to detail from his employees.

In early December, Rusty developed a list of projects for Anne to assist her in preparing for the spring digging season. Her tasks included making an inventory of the nursery's digging supplies. She was to order needed tools and supplies, including digging spades, wire baskets, squares of burlap, and boxes of sisal twine, and to assist Rusty in coordinating service and maintenance of the nursery's trucks, tractors, skid-steers, and other primary pieces of digging equipment. Anne also was to assist Rusty in interviewing and hiring three laborers to fill existing vacancies on the digging crews.

As Anne developed her inventory, Rusty noticed that her listing of tools and related supplies seemed incomplete: only three sizes of burlap were scheduled for order and no digging spades had been added to the order list. Rusty knew the nursery was short by at least four to five digging spades and surmised additional hand tools and supplies were needed. He requested that Anne double-check her inventory list and get back to him with an updated version by the end of that week.

By the following Monday Anne presented a more detailed listing of the nursery's tool and supply inventory. Although Anne's second effort was much more satisfactory than her first, Rusty was still concerned that her initial attempt had been incomplete and that the entire project had taken so long. After he and Anne reviewed the final inventory together, he asked that Anne contact each supplier and have the necessary items shipped by motor freight. After the shipments arrived, Rusty noticed that an inadequate number of boxes of sisal twine and digging spades had been ordered. Although new managers' first performance review was normally scheduled 6 months after hiring, Rusty wanted to give Anne a comprehensive 4-month review to highlight Anne's need for specific improvements before the spring digging season.

In Anne's 4-month review, Rusty noted that she was routinely punctual for their 7:30 a.m. start time, cooperative in working late when needed, and had no unexcused absences. She consistently displayed a very positive outlook and always got along well with other members of the staff. Rusty additionally noted that Anne always communicated clearly, collaborated well with others, was personally accountable for her actions, and adapted well to most professional situations.

Rusty praised Anne for these professional qualities, but he told her she had three main areas of managerial responsibility in which she needed to improve. First, her time-management skills were deficient. Although Anne was always busy, she did not prioritize her time well and too often failed to adequately concentrate on areas of crucial importance. Anne tended to get sidetracked with tasks that were not critical to the nursery's current operation.

The second area of responsibility that Rusty discussed was Anne's inability to complete important tasks in a timely manner. Anne's focus tended to wane, and several times he had had to ask her to double-check her work after her first attempt. Rusty had frequently assigned tasks to Anne a second time. He believed that the more involved or complex the responsibility, the more focus Anne needed to complete the job.

Third, Rusty suggested that Anne concentrate on improving her ability to problem solve in work situations. Although it was not a difficult task to inventory the current number of digging spades, wire baskets, and sisal twine, the consequences of starting the spring digging season with incomplete inventory could be disastrous at such a critical time of year. Anne sincerely appreciated Rusty's constructive comments as they helped her to see her work-related deficiencies. She felt that she was now in a stronger position to solidify her efforts toward preparing for the spring digging season.

Through the critical months of March, April, and May the nursery received a larger-than-usual number of digging orders. A sizable shopping center was being constructed nearby, and a significant percentage of the trees and shrubs specified in the landscape contract were purchased from Farmington. A large, exclusive seniors' community was also nearing completion near the Bay, and the landscape architect overseeing the project purchased many of the needed landscape plants from Farmington. Funding from the state of Maryland had been allocated to replant a significant portion of the greenways that bordered a nearby interstate highway and the contracted landscape architecture firm also used Farmington to supply over 75 percent of the shade and ornamental trees for the project.

Despite a busier than normal spring season, Anne matured significantly in her role as assistant field superintendent at Farmington. She made excellent progress and showed marked improvement in her managerial weaknesses as discussed in her 4-month performance review. Although Anne still displayed some limitations in consistently focusing on the task at hand, she had made measurable improvements in both prioritizing her time and concentrating on tasks of critical importance. Anne worked hard to improve her problem-solving ability, and she had developed into a much more responsible supervisor by the end of the spring digging season.

At the end of May, Rusty scheduled a follow-up performance review to update Anne on her progress as assistant field superintendent. Although she had not yet completed a full year of employment with the nursery, it was clear that Anne had progressed significantly. Rusty specifically noted that Anne had clearly taken a leadership role in problem solving in numerous critical situations. He also was happy that Anne had taken on added responsibility and that she was much more adept at successfully managing multiple priorities.

QUESTIONS FOR DISCUSSION

1. How could Anne Sawyer have made a better transition from being a university student majoring in ornamental horticulture to being assistant field superintendent at Farmington Nurseries?
2. What specifically could Anne have done to more expediently correct her managerial deficiencies during her first 4 months at Farmington?
3. How fortunate was Anne to have such an insightful and proactive supervisor to mentor her maturation as assistant field superintendent with Farmington Nurseries?
4. Rusty Sutter could have reacted in several ways toward Anne's initial shortcomings as his assistant field superintendent. Do you agree that he showed patience and persistence in guiding Anne through her first 4–6 months with the company?

   What do you think of his approach of performing a 4-month performance review to help Anne focus on her managerial deficiencies?
5. How would you have handled the issues associated with Anne's managerial deficiencies if you had been her direct supervisor?
6. Anne displayed many professional attributes, including sound communication skills, an ability to work well and collaborate with others, the integrity to be accountable for her actions, and the fortitude to adapt to most work-related situations.

   If, in the future, Anne had an assistant to supervise, how do you think she might respond to deficiencies the new assistant might have in adjusting to new managerial responsibilities?

CRITIQUE

Anne Sawyer's first full-time job experience, as the assistant field superintendent at Farmington Nurseries, provides several lessons. First, it is clear that Rusty Sutter saw the need for Anne to mature professionally before she could fully realize her managerial potential. It was for this reason that Rusty gave Anne a performance review 2 months ahead of schedule. Although Rusty saw significant potential in Anne's professional capabilities, he also appreciated the need to outline and discuss her work-related deficiencies in an effort to help her mature in her new position.

Anne had a number of positive characteristics. She was a diligent, dedicated, trustworthy, and likable employee. She was professional and worked well with other staff members throughout the organization. But these did not in Rusty's mind fully overshadow her three main shortcomings. First, Anne needed to prioritize more effectively and concentrate on areas of critical importance. Second, Anne needed to focus on and complete important tasks. Last, Anne needed to improve her ability to problem solve and work through critical work-related situations.

When Rusty discussed Anne's managerial shortcomings, he was tactful and used sound judgment in helping her mature as his assistant. Some supervisors might have been much harder on Anne in an effort to make their point, and they might have alienated a diligent, motivated, and talented young assistant. Although Rusty made it clear that it was necessary for Anne to improve, he told her that none of her managerial deficiencies were insurmountable. Anne simply needed to accept her professional weaknesses and dedicate herself to changing them.

For her part, Anne took full responsibility for her actions and did not try to belittle the situation or point blame in another direction. It is also important to appreciate that Anne had showed consistent improvement in rectifying her managerial deficiencies. Anne was an excellent team player. She was also intelligent and possessed the ability to understand and appreciate her managerial faults and correct them through diligence and hard work.

Motivation: Essential to Surviving in the Green Industry

OBJECTIVES

After studying this chapter, you should have an understanding of:

- personalities of staff
- short- and long-term goals
- the right person for the job
- professionally challenge each employee
- developing staff talents
- unified goal
- measurable time frames
- positive reinforcement
- matching the reward to the accomplishment
- performance-oriented team
- honesty, integrity, and trust
- personalizing motivation

Whether planning a career in the landscape, nursery, turf management, or arboriculture profession, being self-motivated is critical. All of these trades are physically demanding and can be extremely seasonal as well. Our Northern states have harsher winters, which shortens the time available for work and makes working efficiently that much more crucial.

Of the attributes that supervisors look for in staff, two are especially key. The first is honesty and the second is motivation. Employees who are untrustworthy and do not have a reasonable work ethic are of little value to the organization. Supervisors who are highly motivated are often much more successful in rising through the ranks of an organization and more likely to have a motivated staff. No matter how impressive a manager's credentials, unmotivated individuals seldom move far up the corporate ladder.

Before we get into the specifics of motivating staff, two additional points are important to consider. First, it is important that supervisors mesh their motivational techniques with the personalities of their staff. What motivates one crew member may be significantly different from what motivates another. We will expand on this later in the chapter. Second, supervisors should consider an employee's short- and long-term goals when personalizing motivation. Just as the professional aspirations of one crew member probably differ from those of another, the motivations of each may be dissimilar. Remember that an employee's work-related goals will often influence his or her level of motivation. ❧

Motivating staff requires that managers

- learn from their past mistakes.
- mentor and support inexperienced employees.
- earn the trust and respect of their subordinates.
- have confidence in their staff's professional abilities.
- retain composure under difficult circumstances.

Seeing the Big Picture

Selection of the right person for the job has to happen before managers can motivate staff. Most supervisors inherit the majority of their staff, but they will need to hire more eventually. After selecting the right person, it is up to the manager to professionally challenge each employee. Staff who are not challenged in their work-related responsibilities have little hope of remaining motivated over the long term. In addition, promoting the professional growth and development of their employees is motivational to supervisors themselves. By successfully developing the talents of their staff, managers can often mold raw but talented personnel into highly productive and dedicated members of the organizational team.

The story of Strawbridge Landscape Services illustrates how employee motivation can affect a company. Strawbridge is a good-size design and installation firm well into its second generation of ownership. The company was started over 30 years ago by John Strawbridge Sr. and is now additionally managed by his two sons, John Jr. and Mark. The Strawbridges received a large and lucrative contract to design and install new landscape plantings around the corporate headquarters of Sorenson-Meyers Industries. Sorenson-Meyers is one of the primary providers of optical and diagnostic instrumentation for the medical industry. It was initially agreed that John Sr. and his youngest son, Mark, would assume primary oversight of the Sorenson-Meyers project. John Jr. would manage Strawbridge's other landscape accounts. Within his first 2 weeks at the job site, Mark Strawbridge was 30 minutes or more late for work on three occasions. He made no effort to interact with any of the full-time crew members and gave little respect to the landscape foremen. Because of the high personal integrity of John Sr. the company had had few personnel-related problems throughout its history. Many long-time employees had served for more than 20 years.

Mark had been involved in the hiring of nine seasonal laborers, and he had not checked their background information before their hire. Three of them quit in their 1st week of employment and two others had to be terminated by John Strawbridge Sr. for failure to follow directives from their immediate supervisor. Because Mark made virtually no effort to become involved in the planning process, he often scheduled work that had limited importance to the overall completion of the project. On one occasion, he directed a crew of three full-time employees to install plant material for most of the day in a location of only minimal importance. In addition to Mark's inability to work toward long-term goals associated with the project, he was given little respect from full-time members of the design or installation teams.

Strawbridge Landscape Services successfully completed the landscape project at the corporate headquarters of Sorenson-Meyers Industries, finishing ahead of schedule. But it had been a managerial nightmare for the company because Mark Strawbridge failed to appreciate that his professional behavior was detrimental toward motivating other staff associated with the project. He overlooked the importance of choosing quality employees and failed to mesh motivational techniques with the personalities of his staff. Mark Strawbridge also ignored project planning. His motivational and leadership qualities were seriously deficient and did nothing to foster the long-term goals of the organization. Any success associated with the project was due to the exceptional efforts of John Strawbridge Sr. and his eldest son John Jr., along with many of the company's long-dedicated staff.

Supervisory qualities that help motivate staff include

1. being respectful.
2. being confident without being arrogant.
3. giving credit to staff for their accomplishments.
4. managing with consistency.
5. working through difficult situations.
6. focusing on long-term goals.
7. being trustworthy and ethical.

Developing a Plan

When supervisors fail to develop a plan and are unable to establish meaningful parameters in the supervision of their staff, most often it is because they did not break the motivational process into a manageable progression of steps. The astute manager looks at the big picture. He or she then directs staff in their day-to-day responsibilities according to the long-range goals of the organization. Such an approach allows staff to more personally identify with the jobs or tasks at hand. Knowing how the organizational puzzle fits together allows staff to work together toward a unified goal. In addition to managing staff toward a common goal, employees must be directed to work within measurable time frames. Although most supervisors are effective in prioritizing goals for their staff, they are much less reliable in assigning quantifiable timeframes toward the completion of important jobs or tasks.

Almost any employment psychologist would argue that positive reinforcement is a valued tool in proactively motivating one's staff. Matching the reward to the accomplishment is part of positive reinforcement. A supervisor would not reward a seasonal intern working on a landscape crew with a new luxury automobile for successfully planting a small tree, an absurd relationship between achievement and reward. Let's imagine that a key manager in charge of a landscape firm's Design & Installation Division and an equally senior supervisor within the company's Landscape Maintenance Division left to jointly start their own business. Unfortunately, they terminated their relationship with their employer in March—right at the start of the busy spring season. The general manager worked long days, nights, and Saturdays to pull the company through until suitable replacement personnel could be found. The general manager did an outstanding job meeting responsibilities to clients and preventing tens of thousands of dollars in revenue being lost. If the company's owner had showed her gratitude by providing the general manager with a new pair of hand pruners, such a puny act of generosity would have been more insulting to this extraordinarily hard-working senior staff member than no acknowledgment at all. Pruners would have been as ridiculous in this situation as the luxury automobile for planting a small tree in the preceding one. Although these two scenarios are obviously extreme; it is vitally important for supervisors to closely coordinate the appropriate reward with the accomplishment.

It is also important to choose rewards with the interests and personality of the employee in question. A day's fishing trip on an ocean charter will be worthless to an employee who is bored to tears by fishing or gets seasick with the first moderately high wave. In addition to matching the reward to the accomplishment, recognize accomplishments within a reasonable time span. Thank-you parties and similar recognition events for university interns are almost always planned for late summer, just before the students return to college. For full-time staff, the timely recognition of accomplishments should be no less important. Although it is tempting to wait for a slow period in the business cycle, waiting too long can lessen the sincerity of the reward.

Motivating supervisors are those who

- acknowledge the positive accomplishments of their staff.
- support cooperation among their staff.
- maintain expectations of quality work.
- expect staff to be self-disciplined.
- have professional ethics and integrity.
- believe in a team approach.
- mentor less experienced members of the team.

Understanding the Needs of Others

For many employees, a sense of accomplishment and knowledge that they achieved their personal best are the motivators. Some like to have their ego stroked. A sense of challenge can be motivational for those with an adventurous side; the exclusivity of being selected as a member of an elite group can be highly motivational for others. A supervisor who understands the motivational needs of each employee is a supervisor who develops a performance-oriented team that works in concert with the long-term goals of the institution.

Specific motivators are universally important to nearly all employees. Rare is the employee who does not want to be appreciated by his or her supervisor. Respect from peers is another powerful motivator in the workplace. Most employees find emotional fulfillment, another motivator, in having their accomplishments recognized, even if it comes in the form of a simple thank-you. The core values of honesty, integrity, and trust are three critical motivators that belong in all work environments. Quality employees expect guidance and discipline in the workplace. Being involved with a company that has such an atmosphere fosters, supports, and motivates staff at all levels.

Notice that motivators have not been rated or prioritized: all can contribute to a quality work environment. Notice also that money was not included. Money is important, but it is often not the chief motivator for many employees.

Unlocking the Motivational Door

To motivate employees, supervisors must understand their innate behavior. The problem is that everyone is different. No two employees react to problems or stresses in the same way. Neither do they plan their schedules or structure their time in the same manner. They seldom communicate in a similar manner and do not problem solve or relate to others in the same way. Supervisors will not always be able to appreciate why certain staff members think or act as they do. This can be frustrating for supervisors and make it hard for them to maintain their patience and understanding in the day-to-day supervision of their staff. However, managers who try to understand their staff have a better chance of being trusted and effective motivators.

Staff are motivated when they know their supervisor

1. is consistent.
2. can handle stress associated with his or her position.
3. is able to plan and provide direction.
4. can communicate his or her objectives.
5. can clearly delineate tasks and responsibilities.
6. encourages staff input.

Motivating Difficult Employees

Regardless of their best efforts, supervisors will never be able to successfully motivate every member of their staff. Some people resist any attempt to motivate them and go out of their way to be negative about everything in their lives. If you took one of these people to a fine restaurant and encouraged them to order anything on the menu, they still could not say anything positive about the experience. For others, continuity and avoiding change is extremely important. You have probably worked with someone who brought a ham and cheese sandwich to work every day, had to use the same shovel week in and week out, or had some other unvarying behavior. In most of these cases, the choice of sandwich or shovel is not a decision based on need but a decision to avoid change. Then there are the employees who are overly protective of their job function—a certain area of responsibility must remain exclusively theirs. Early in my career I worked with a mechanic who oversaw equipment maintenance for a large landscape company. The mechanic would not allow other employees to even borrow a screwdriver to use in his presence. It was not a matter of damaging an important tool. The mechanic believed that only he was capable of administering to the equipment-related needs of the company. He was also extremely protective of his workspace and was quick to let trespassers know what portion of the shop was his domain. In another position I worked indirectly with a supervisor who would not give his assistant keys to locked equipment cabinets. He was not worried that chain saws or hedge trimmers would be stolen; the supervisor was being ultraprotective of his professional responsibilities.

For countless reasons you will never be able to motivate a small percentage of your staff. Some employees thrive on being negative and oppose any positive change in the workplace. Other employees resist attempts to motivate them because you are of the opposite sex, much younger, or a different ethnicity. Still others hold a grudge against you because you were promoted to the position they wanted or because you have a college education. As ridiculous as it sounds, some employees resist attempts to motivate them because you drive a brand of pickup truck they don't like.

Personalizing Motivation

One of the first steps toward personalizing motivation, or finding what motivates an employee, is to gain his or her trust. You can never hope to gain employees' trust if you are not trustworthy. If you expect employees to have confidence in you as a leader, you must be loyal to them and they must believe that they can depend on you. You must be respectful of your subordinates. This doesn't mean that you cannot discipline your staff; it means you should do so with civility. It is also important to be consistent in managing your staff and not play favorites. Do not second-guess your staff. If you gave clear directives and have confidence that they can perform professionally, allow them to do their job and show their professionalism. Work hard to give your staff the latitude to follow through on their responsibilities. This will go a long way toward fostering a trusting environment between you and those you supervise.

Everyone wants to be accepted. Employees appreciate having their opinions heard and respected. One of the biggest motivators for many employees is to allow them ownership in the decision-making process. Easy to comprehend, this concept is frequently underused as a motivator. In the traditional top-down style of management, supervisory personnel establish the tone and direction of the company and lower-level staff carry out assigned responsibilities. Depending on the experience and foresight of senior management, this can be an efficient method for operating a profit-oriented business. But it can also frustrate capable young employees by stifling their fresh ideas. In contrast, encouraging subordinates to be more involved in their job and share ideas fosters commitment. In this more participatory style of management, upper-level supervisors set the focus and direction of the organization and take responsibility for its ultimate success or failure. Allowing staff to accept ownership of the day-to-day decision making and promoting their voice in shaping the long-range goals of the organization will translate into more productive, responsible and motivated employees.

Motivation in the Green Industry

A recent survey conducted by Lee Marcus, director of human resources at Marcus Farms, a wholesale nursery, asked the nursery's employees to rank the motivating factors associated with their jobs (Marcus, 2007). The survey was based on initial research conducted by James Lindner, who asked workers, many of whom were horticultural workers, to rank the following motivators:

- job security
- sympathetic help with personal problems

- personal loyalty to employees
- interesting work
- good working conditions
- tactful discipline
- good wages
- promotions and growth in the organization
- feeling of being in on things
- full appreciation of work done

Lindner found that the top four job-related motivators in order of preference were (1) interesting work, (2) good wages, (3) full appreciation of work done, and (4) job security. Other than modifying Lindner's "good wages" to "good wages/benefits/bonuses," the list used by Marcus was exactly the same. Marcus found the top 10 motivators for employees at Marcus Farms to rank slightly differently, the first four being (1) job security, (2) good wages/benefits/bonuses, (3) full appreciation of work done, and (4) interesting work (Marcus, 2007, pp. 32–35).

Although wages and overall job-related compensation ranked high, note that money was not the top motivating factor in either survey. Job environments that stimulate employee interest and promote the feeling that employee efforts are appreciated received high ranking in both surveys. Perhaps the top ranking that employees at Marcus Farms gave job security was because a high percentage of seasonal workers were surveyed, or maybe the local economy was depressed and employees had few other employment options. The overall results do imply, however, that managers who foster a team approach and include workers in the decision-making process will more successfully motivate staff. Both surveys' results suggest that listening to employee opinions and creatively managing assigned tasks do much to promote a positive and productive work atmosphere.

In her conclusions Marcus stressed the importance of developing a reward strategy for positively motivating employees. Spontaneity in staff recognition coupled with personalizing rewards to individual employees yield large returns. In addition, recognition of noteworthy accomplishments must be sincere. Finally, Marcus observed that supervisors must support progress toward long-term, incremental change and anticipate problems. Quality motivational programs take time. They are, however, well worth the effort (Marcus, 2007, pp. 32–35).

CONCLUSION

Efforts to motivate staff must coordinate with the overall management philosophy of the organization. Motivational strategies should also be coordinated into a unifying thread throughout the company. Developing a keen understanding of the employees you manage helps you unlock their potential, which helps them in making well-thought through and proactive decisions. Remember, that the fostering of motivated and responsible employees is of paramount importance to the long-range goals of any organization.

DISCUSSION QUESTIONS

1. Why should managers develop the professional abilities and talents of their staff?

   How does this practice contribute to the long-term health of horticultural organizations?
2. Why should motivation and honesty be considered when evaluating an employee's value to an organization?

   If you had to choose only one of these two character traits, which would be most important to you as a supervisor?
3. Give some examples of a manager's level of motivation being contagious to other members of his or her staff.
4. Why is personalizing motivational techniques a critical component of getting the most productivity from employees?
5. If you were John Strawbridge Sr., how would you handle the lack of growth and maturation of your son Mark, especially as related to becoming a productive, long-term partner in the family business?
6. If you were John Jr., what expectations would you have for your younger brother's maturation as a future partner in the family business?
7. If you were Mark Strawbridge, what three primary deficiencies would you have to overcome in order to become a trusted and responsible partner in the family business?
8. In your experience have most supervisors been diligent in assigning time frames for completing responsibilities?
9. How important would it be for you as an employee to receive positive feedback from your supervisor for an above-average effort?

   Do you believe that most staff respond favorably to positive reinforcement from their supervisors?
10. How important is it for supervisors to consider the overall personality and emotional makeup of a staff member when considering recognition awards?
11. A sampling of staff motivators might include (1) provision of respect, honesty, and integrity; (2) professional fulfillment; (3) recognition for excellence; (4) wages and overall job compensation; and (5) direction and discipline in the workplace.

    What would you add to this list?

    Rank items in this list in terms of what matters to you.

12. Do you agree that being overly protective of job duties can cause some employees to resist being motivated by their supervisor?
13. The importance of a supervisor being honest and ethical was discussed in relation to motivating his or her staff. It was also stressed that supervisors should try not to second-guess the professionalism of qualified staff.

    Do you agree with both of these managerial concepts?

    Are there any additional trust builders that you believe are critical to motivating staff?
14. How important is it for motivational strategies to be closely tied to the long-range goals of the organization?

Sarah Chen's Supervisory Experience at Stone Harbor Landscape & Design

Sarah Chen graduated first in her class of Landscape Architecture students and also achieved a minor in Ornamental Horticulture. Sarah's long-range professional goal was to own her own Landscape Architecture business, but she knew she first needed practical, hands-on experience working with a quality landscape installation company. Sarah decided to accept the position of landscape supervisor with Stone Harbor Landscape & Design. She believed that wider professional experience would benefit her career goal to be a registered landscape architect.

Sarah was extremely self-motivated and quickly gained the respect of her staff. Of the two landscape crews that Sarah had responsibility for, she could readily tell that one of her crew leaders resisted her as his supervisor. John Bates had been a crew leader and part-time mechanic with Stone Harbor for the last 4 years. John had hard feelings toward Sarah for two main reasons. Stone Harbor had considered John for the position of landscape supervisor, but dyslexia made him a slow reader and writer. John knew he could do the job and resented Sarah for getting the position he wanted. Deep down, he also resented her for having the college degree that circumstances had denied him. John believed that through hard work he could have secured a baccalaureate degree, despite his dyslexia, if he had had the opportunity.

Although John was not outwardly disrespectful to Sarah, it was clear he was not pleased that she had been hired as his supervisor. Just a couple of weeks after Sarah's hiring, she was reviewing the progress John's crew had made in installing a large, expensive residential landscape. John's crew was also completing some minor arboriculture work. He told Sarah he could not start the chain saw and handed it to her. Sarah quickly noticed that John had disconnected the spark plug wire from the top of the plug, disarming the saw's ignition system. She casually reconnected it, saying that the saw should now start.

About a week later, John was on another installation job and wanted the front-end loader moved closer to the dump truck. Sarah happened to be on site, and instead of asking one of his crew members to reposition the loader, he asked Sarah to move it. He made his request in front of two full-time crew members and a seasonal intern, thinking that Sarah would be embarrassed by not knowing how to operate a loader. To his surprise, Sarah jumped into the machine's seat, immediately started the machine, shifted into reverse, and deftly moved the loader to the opposite side of the dump truck. She then checked with John to see if he needed any additional assistance before she left to check on her second crew.

The next day, John asked one of his interns to tidy up a grass edge with a gas-powered string trimmer. As the employee was unloading the string trimmer from the equipment trailer, John asked Sarah if she would assist the intern with fueling the machine. Sarah obliged and showed the new employee how to start and operate the machine. She also made sure that he was wearing his eye and ear protection. She then stood by for a few minutes to be assured that he could safely operate the machine.

About a week later, John's crew was on another job when Sarah arrived. John asked if they could sit and talk for a few moments on a nearby bench. She agreed, and John apologized for his behavior over the last few weeks. He said he was extremely impressed with her overall horticultural knowledge and familiarity with the operation and maintenance of horticultural equipment. He also said he was favorably impressed by the professional manner she used when she supervised and the respect she showed her staff. Sarah thanked John for his comments and asked that they develop a closer working relationship. She said that she had several ideas on how the crew could work more efficiently but would value John's input in further developing the crew.

Sarah worked for a little more than a year with the firm before moving on to work with a well-known landscape architecture firm in the Midwest. Before she left, she recommended that John succeed her in the position of landscape supervisor.

QUESTIONS FOR DISCUSSION

1. Put yourself in the position of the owner of Stone Harbor Landscape & Design. If Sarah had not applied, would you have promoted John to fill the position of landscape supervisor with responsibility over two crews?
2. Do you agree with Sarah's decision to gain hands-on practical experience for a year before furthering her career as a landscape architect?

3. Why do you think Sarah had not mentioned her equipment-related knowledge to her staff?

   Do you think John would still have tested Sarah in her new position?

   Should Sarah have expected John to test her knowledge of horticultural equipment?

4. If you were Sarah, would you have handled your supervisory experiences with John any differently?

5. Do you agree with the owner's decision to hire Sarah as landscape supervisor with management responsibility over two crews?

   Was Sarah ultimately successful in her new position?

6. How many young supervisors would have been able to troubleshoot the equipment challenges that Sarah correctly diagnosed?

   Is it necessary for a horticultural supervisor to be able to perform the majority of tasks that his or her staff do?

   How important is it for a young supervisor to be reasonably familiar with the overall operation and maintenance of horticultural equipment?

7. Imagine you are the owner of Stone Harbor Landscape & Design.

   If after hiring Sarah as a supervisor over two landscape crews you found that she was not experienced enough to handle the position, what steps would you take to remedy the situation?

   Depending on the issues in question, how long would you work with Sarah if you saw that she still held managerial promise to fulfill the responsibilities of the position?

8. Do you have any doubts that Sarah would be successful in her future role as a landscape architect?

   What personality traits do you think Sarah exhibited that would make her successful as a future Green Industry entrepreneur?

CRITIQUE

John and Sarah developed a close working relationship before Sarah moved on the following year to work for a landscape architecture firm in the Midwest. Before Sarah left the company, she met with the owner of Stone Harbor Landscape & Design to recommend John as her replacement for the position of landscape supervisor. Why did this initially prickly relationship develop into a positive one and not become a managerial nightmare for Sarah? In answering this question, remember that Sarah's goal was to gain a year of practical, hands-on experience before pursuing her future as a landscape architect. Most in Sarah's situation would not have been so far-sighted as to first walk in the shoes of those who would be installing her future designs.

Although John knew that Sarah had graduated with a degree in landscape architecture, he did not realize that her Chinese father was a veterinarian who specialized in livestock and that her Pennsylvania Dutch mother had been brought up on a dairy farm in Lancaster County, Pennsylvania. As a result, Sarah spent most summers throughout her middle and high school years running machinery and helping with livestock-related chores on her uncle's and grandfather's farms. It was through Sarah's rural upbringing that she became skilled with the operation of many types of farm-related equipment.

Another reason for Sarah's success was that she took time to understand and evaluate the personalities of her crew members. She knew that John did not have a great deal of professional respect for her when she began working at Stone Harbor. John did not realize that Sarah was working entirely within her comfort zone and that operating farm-related equipment was second nature to her. Importantly, Sarah was able to handle each situation that John presented her and did not show anger and berate him in front of his crew. As a result, there were no axes to grind on either side. In Sarah's own subtle yet confident way, she let John know she was an adept and qualified supervisor. Sarah did not have to be rude or disrespectful to get her point across and she worked each situation through with commendable results.

Sarah was in control of each situation and mature in the way she handled herself as a supervisor. Sarah's equipment-related abilities were readily apparent to the owner of Stone Harbor when she was hired for the position of landscape supervisor. It was not normal hiring procedure for the company to give a recent college graduate authority over two landscape crews. However, through the interview process, the owner knew that Sarah was especially mature and would be well able to handle the responsibilities of the position.

The owner of Stone Harbor could see that Sarah was not going to be a long-term employee with the firm. He was aware, however, that Sarah was bright and capable and that she would be an excellent short-term addition to the company. Few owners would have hired Sarah, even though truly exceptional people are hard to find. Far-sighted managers hire bright and capable employees despite knowing that they will be heading off to fulfill their dreams in the relatively near future.

Finally, it was admirable for John to rectify his working relationship with Sarah. John benefited greatly from acknowledging that Sarah was an extremely capable supervisor. It was insightful and mature for Sarah to want to bring John into the fold and include him in the decision-making process. The experience enabled John to be considered for the position of landscape supervisor when Sarah left the company.

Decision Making: Making or Breaking Your Future with the Company

After studying this chapter, you should have an understanding of:

- long-term goals
- word-of-mouth advertising
- melting pot of cultures and ethnicities
- long-satisfied customers
- contingency plan
- fluid and malleable in decision making

In addition to time management and motivation, discussed in previous chapters, decision making is important to the career of a young supervisor in the Green Industry. ❧

Decision Making and Long-term Goals

Successful horticultural companies coordinate daily decision making with the organization's long-term goals. This is as applicable to an arboriculture firm in northwestern Pennsylvania as it is to a commercial orange grove in south-central Florida, to a small privately owned garden center as it is to a large and rapidly expanding commercial greenhouse facility. The ultimate success of any horticultural operation depends on its ability to formulate and execute day-to-day decisions in accordance with long-term goals.

A study of decision making by business entrepreneurs attempted to uncover "the strategic, or turning-point, decisions in [entrepreneurs'] business careers" (Ehringer, 1995, p. 2). An entrepreneur interviewed for the study said:

> One of the things that I keep emphasizing over and over again is why you have to work day-to-day and make decisions in terms of the long run. I've always said to everyone all the time . . . don't ever make a

> short-term decision, make your decisions based on something you can live with a year from now or two and feel good about. You don't always have it turn out that way but that's got to be the guiding principle. (Ehringer, 1995, pp. 54–55)

The overall philosophy of an organization closely mirrors that of its senior leadership. Owners in the Green Industry should strive to never have their professional ethics questioned. They should be known throughout the industry as honorable entrepreneurs and admirable role models in the local business community. Treating all customers with honesty and integrity will earn the company loyal customers who recommend it to others. Business in the Green Industry is based on word-of-mouth advertising and the repeat business generated by long-satisfied customers.

As the entrepreneur in the study said:

> I am absolutely sure . . . that the tenor of the organization will come from the top. And you can't fake it. If you get to know the person who's running the organization, you'll have a pretty good feel for the organization . . . it does permeate somehow . . . business and the decision-making process. (Ehringer, 1995, pp. 54–55)

Staff resistance to a supervisory decision is often the result of

1. their confusing it with other, unrelated critical issues.
2. supervisory failure to listen to and respect their concerns.
3. their insufficient acceptance of the overall process.
4. lack of communication between supervisor and subordinates.
5. a misunderstanding of delegated assignments or responsibilities.

Decision making must remain ethical. This cannot be stressed enough, regardless of whether we are talking about short-term responsibilities or long-term goals. A young supervisor who finds that company policy compromises his or her professional ethics needs to strongly consider going elsewhere.

Looking to the Future

Decisions made today can impact a landscape, nursery, arboriculture, or turfcare company weeks, months, or even years into the future. Decision making at any level must take into account not only the current short- and long-term goals of the company but the flexibility of goals, which may need to change as the organization grows and matures over time.

A component of sound decision making is the contingency plan. Many a mature and seasoned manager has had to learn the hard way that a back-up plan can spell the difference between success and failure. A contingency plan presents a viable alternative, always needed in the highly seasonal horticulture field.

Keep an open mind toward all aspects of the decision-making process. Objectivity is an important part of decision making and a trait even mature managers too often fail to cultivate. Objectivity and the ability to develop and maintain a quality working relationship with individuals different from you, of varied cultures, ages, and backgrounds, will help you deal with the huge melting pot of cultures and ethnicities in the Green Industry.

The Benefits of a Mentor

There is a significant difference between managing only your own productivity and managing the day-to-day activities of others. Learning to make the right decision and being able to do so in a reasonable time is a very important part of being a truly effective manager. It is as important for supervisors to make timely decisions as it is to not make decisions too quickly. More often, however, young supervisors delay making decisions, waiting for a more opportune time that usually never occurs. Unfortunately, making managerial decisions is not something learned from a textbook. Real work-world experience is the best teacher for giving young supervisors confidence to make decisions. In any decision-making situation, young supervisors must understand the importance of effectively managing staff. There is a significant difference between being responsible only for your own productivity and managing the day-to-day activities of others. As a result, the recent graduate who is able to make the distinction between personally completing a job or task and managing the work-related responsibilities of an entire crew will be infinitely more successful to the organization.

Being mentored by a well-seasoned manager early in your career will help you develop the talent for managing staff. Many topflight managers would acknowledge that they were fortunate in having a mentor early in their career. Role models are priceless and can make a young supervisor's climb up the managerial ladder much less tempestuous. A mentor should be someone you trust and admire as a professional in the Green Industry. To help you select a mentor, ask yourself whether he or she makes mature and well-thought-through decisions. Does this person have high ethical standards? Does he or she lead with a confident yet cordial demeanor? Is he or she clear in communicating goals and objectives to staff? Is this person fair in managerial expectations, and is staff willing to overachieve because of a mutual respect and admiration? Does he or she motivate staff, and do subordinates believe their hard-earned accomplishments are truly appreciated? Do staff believe their concerns are respected and fairly evaluated? These are only some examples of questions you should ask.

If the decision-making process has not gone smoothly, consider

1. whether you as a supervisor have deviated significantly from the agreed-upon plan.
2. redefining or reorganizing your managerial efforts in an attempt to move forward.
3. seeking the guidance and advice of a mature and trusted manager.
4. whether you adequately understand all of the issues involved.
5. whether you are fairly evaluating the overall situation.
6. whether there has been sufficient effort to communicate with appropriate staff.
7. whether staff have helped in problem solving any deficiencies in the plan.
8. whether experienced and mature staff have been sufficiently involved.

Informed Decision Making

Decision making sometimes must occur amid an ever-changing and complex array of circumstances. To develop a reliable decision-making process for all circumstances, invest adequate time in making decisions. Problem solving requires sufficient hours, days, or more to arrive at a solution; a rash decision could have disastrous effects in some situations. Evaluate a range of options, considering all alternatives.

Remain fluid and malleable in decision making. Decisions made last week, last month, or last year might be totally inadequate for current problems.

Information can become outdated and decisions made based on that information can now be ill-suited to the current needs of the organization. Managers must use the best and most current information available to formulate decisions.

Consider the pros and cons of each decision. Supervisors who fail to think through the potential implications of a decision may be plagued by unpleasant consequences. Suppose you are a supervisor and your crew laid 5,000 sq ft of sod the morning of a hot August day forecast to peak at 97°F. and fall only to 89°F. during the night. You and your crew know that sod dies in less than 24 hours if it dries out. You instruct one of your subordinates, Jim, to irrigate the sod before he leaves at the end of the day. Jim helped install the sod and knows it requires a thorough soaking in the current weather conditions. You leave and check various jobs all afternoon, with your last stop being the sod installation. You find Jim gone and no sprinklers set up. You now have some decisions to make.

The watering system does not have a timer, and you cannot leave the irrigation system operating throughout the night. The forecast is for the weather to break tomorrow, with a high of only 90°F. But will the sod live until tomorrow morning? Is it wise to take the chance? The sod was installed in the front yard of Mr. and Mrs. Duran's new $2,250,000 home in a new, exclusive neighborhood that could grant similarly large and lucrative landscape contracts from other homeowners. The name of the landscape firm has been prominently displayed on both doors of your company pickup truck all through the installation.

Is there any choice but to turn on the sprinklers and return in a few hours, despite the round trip being more than an hour's drive, to shut the system down? Are you willing to risk your employer's reputation and your job to avoid the drive? From almost any perspective the answer is quite simple, as 5,000 sq ft of dead sod in front of the Duran's new residence is not the advertising a landscape company seeks.

The Consequences of Risk-taking

Let's revisit the endangered sod at the Duran residence and make a few changes to the risk calculation. Although it is still true that Jim failed to water the sod, its square footage has shrunk to less than 600 sq ft. It was not laid at the Duran's new residence in an exclusive neighborhood but at the very modest home of Mr. and Mrs. Conigliaro. Although the Conigliaros live in a comfortable middle-class neighborhood, their home is worth a fraction of the $2,250,000 the Durans paid for their property, and this neighborhood offers little opportunity for lucrative contracts for a landscape company.

These circumstances dramatically change the situation. It has been a long, hot day for everyone. Is it worth more than an hour's drive to ensure the sod lives until morning? The risk of being terminated for failure to water less than 600 sq ft of sod in a modest, middle-class neighborhood seems negligible. Perhaps the worst outcome would be a stern verbal reprimand from your supervisor should the sod not make it through the night.

Did you forget about being responsible for your actions? Did you forget about the potential consequences of rash decision making? Did you forget that the owner of the landscape company you work for is Mrs. Conigliaro's brother? Isn't it interesting how relationships influence a decision?

The following are valuable for assessing decision-making capabilities:

1. Do you make decisions in a reasonable time or do you procrastinate?
2. Do you delegate a sufficient degree of decision making to your staff?
3. Do you evaluate all pertinent information before making a decision?
4. Are you analytic when making decisions or do you evaluate situations emotionally?
5. Do you try to serve both the long- and the short-term goals of the organization when making decisions?
6. Do you seek guidance from others when making critical decisions?
7. Are you creative in your decision making?

Making Tough Decisions

A timely and careful decision is of no value if you cannot get key staff to support its eventual implementation. A young supervisor might have to convince senior management of an important decision's value to the organization. Management will want to see that the decision making has considered the overall needs of the organization. When garnering support for a critical decision, young managers should not circumvent their direct supervisor in the process. Supervisors who are clear in communicating the focus and intent of the decision will be more persuasive. Finally, never forget that decisions made today impact a company next week, next month, and well into the future.

CONCLUSION

When making decisions, establish a contingency plan, remain ethical, and maintain an open mind. Consider the consequences of all decisions. Understand company politics and mesh short- and long-term goals in the decision-making process. Look for a mentor who will accelerate your learning. Coordinate an important decision with relevant managers.

DISCUSSION QUESTIONS

1. Give three examples of how a critical decision made today can affect a landscape, nursery, or turf-related company years later.
2. Fully explain why intelligent decision making should be coordinated with both the short- and the long-term goals of an organization.
3. Analyze why a contingency plan is vital to the day-to-day responsibilities of managers in the Green Industry.
4. What could you as a young supervisor gain from the mentoring of an older, mature manager who was long on experience but short on formal education?

5. With respect to the melting pot of cultures and ethnicities which is so representative of the Green Industry, how important is it for young supervisors to consider differing cultures and ethnicities in their decision making?

   How does the fluency of one's Spanish affect a young supervisor's ability to be an effective decision maker?
6. How would you as a young supervisor have handled the issue of irrigating the newly laid sod at the Duran residence?
7. If Jim had no excuse for failing to water the sod, how would you have handled the situation as his direct supervisor?

   What disciplinary measures would you have recommended for Jim?

   Would it be wise to consult your immediate supervisor regarding any disciplinary measures you planned for Jim.
8. Suppose Jim had taken to the hospital a coworker who had cut himself with a pruning saw and required 12 stitches.

   Jim planned to water the sod after his trip to the hospital but was detained in the emergency waiting room for 3 hours. He returned to the Duran residence at 8:00 p.m. to water the sod but saw that you had just left after turning off the sprinklers. As a young supervisor, how would you have handled this situation?
9. How critical is it for supervisors in this age of technology to be able to communicate with his or her staff? Would a company cell phone or radio issued to the crew leader have provided a partial answer to Jim's dilemma at the Duran residence?

Jonathon Hough's Managerial Experience at Nguyen's Garden Center

Jonathon Hough had graduated from a small technical college in a mid-Atlantic state and was now completing his first year as assistant nursery foreman with Nguyen's Garden Center in Streibert, Massachusetts. Jonathon had not been an outstanding horticulture student in college and graduated with a 2.72 overall GPA. Although Jonathon did have some minor learning disabilities, the main reason he did not achieve higher grades was due to a distinct lack of motivation. Jonathon excelled only when he was operating skid-steers, front-end loaders, chain saws, and related equipment, which he did as part of the college's strong technical curriculum.

Mr. Nguyen understood that Jonathon had not been a stellar academic student but believed that he would be able to motivate his new assistant nursery foreman with the nursery's large number of hands-on horticultural tasks and frequent opportunities to operate a wide array of equipment. Mr. Nguyen had hopes that Jonathon would develop into a dependable full-time employee when he first hired him in March. However, Mr. Nguyen started to have reservations during Jonathon's first few months of employment. Jonathon needed to be directed through nearly all tasks. By midsummer it was clear that Jonathon's lackadaisical attitude accounted for his work-related deficiencies and not, as Mr. Nguyen had hoped, his uncertainty in his new role.

Mr. Nguyen was an ethical employer, with the majority of his employees having been with the firm for a minimum of 15 years. Mr. Nguyen had owned and operated the garden center for nearly 28 years and had established a reputation for providing high-quality service and merchandise. His garden center was well-known for its varied selection of trees, shrubs, and herbaceous perennials. It sold a wide range of seasonal items, including annual flowers, vegetable plants, seeds, bales of straw, pumpkins, mums, and Christmas trees. The garden center had an extremely dedicated clientele.

Mr. and Mrs. Nguyen's only daughter was getting married in November, and they decided to take a week off and fly to Seattle. During the week before they left, Mr. Nguyen sat down with Jonathon in two lengthy meetings and outlined what needed to be accomplished while he was away. Throughout his time in Seattle, Mr. Nguyen wanted Jonathon and three seasonal crew members to provide winter protection for the woody and herbaceous plant material throughout the outside nursery area of the garden center. Jonathon was to use the front-end loader and heel-in specific B&B material with bark chips. He was to winterize the lathe house with plastic, change both layers of plastic on the large polyhouse, and move all of the potted herbaceous perennials into the four large cold frames. Mr. Nguyen had always been vigilant about keeping first-class stock at the nursery, and Jonathon was likewise to cull nonsalable plants.

Mr. Nguyen detailed everything he wanted Jonathon to accomplish and made sure that he fully understood his responsibilities. Mr. Nguyen left the telephone number of the hotel where he and Mrs. Nguyen would be staying. Mr. Nguyen clearly told Jonathon that he was to concentrate solely on the list of responsibilities they had discussed and was especially specific as to each job's priority. He believed his leaving for a week would be a good test to see whether Jonathon could be a responsible member of the staff. Mr. Nguyen had furthermore decided that if Jonathon's performance was not adequate during his week on his own, he would not be retained during the winter or rehired in the spring.

Working with his seasonal crew, Jonathon was diligent in using the large pile of bark chips near the lathe house to heel-in trees and shrubs throughout the nursery. However, he did not finish the mulching process with the second large pile of bark chips at the back of the garden center. Jonathon and his crew moved herbaceous perennials to the cold frame but in doing so jumbled together cultivars of *Phlox paniculata*, *Liatris spicata*, *Coreopsis verticillata*, and

Platycodon grandiflorus. Jonathon covered the polyhouse with two layers of plastic, but that was a lower priority than heeling-in the remaining trees and shrubs throughout the nursery, which he failed to do. What's more, excess plastic was not neatly trimmed nor were enough fasteners used to adequately secure the plastic to the frame. For the most part, the quality of workmanship displayed in the covering of the polyhouse was far from acceptable and part would need redoing when Mr. Nguyen returned from Seattle.

Upon returning, Mr. Nguyen saw that things had not gone according to his plan during his week in Seattle. It was readily apparent to Mr. Nguyen that the quantity of work completed was low and so was the quality. Mr. Nguyen summoned Jonathon to his office, where his young assistant nursery foreman was quick to apologize for his obvious lack of motivation throughout the previous week. Jonathon acknowledged that he lacked discipline in his life and said he was seriously investigating joining the Coast Guard. He gave Mr. Nguyen notice that he would be leaving the company and sincerely thanked the owner for the opportunity he had given him. Jonathon stayed on with the firm for an additional week, and then left to join the Coast Guard.

QUESTIONS FOR DISCUSSION

1. Knowing that Jonathon's overall academic achievement in college had been mediocre, was it reasonable for Mr. Nguyen to expect that Jonathon would mature into an acceptable full-time employee?
2. Name five professional qualities that Mr. Nguyen should have expected from his assistant nursery foreman.

   Name five personal qualities that Mr. Nguyen should have expected from his assistant nursery foreman.

   How many of these 10 personal and professional qualities did Jonathon possess?
3. Mr. Nguyen had been clear in communicating the responsibilities he expected Jonathon to complete during his week away in Seattle. Were Jonathon's assigned tasks fair and reasonable?
4. Was there any reason why Jonathon and his crew could not have kept the herbaceous perennial cultivars properly organized in the cold frame?

   Was there any reason why Jonathon and his crew could not have neatly covered the polyhouse?
5. Did you expect that Jonathon's employment would be immediately terminated during their meeting in Mr. Nguyen's office?

   Why did Mr. Nguyen allow Jonathon to finish out a final week of employment before leaving for the Coast Guard?

6. Name five specific changes you believe Jonathon needed to make before he could become a diligent and responsible supervisor.

   Do you think that Jonathon's enlisting in the Coast Guard would support his obvious need for increased discipline in his professional life?

7. Was Mr. Nguyen remiss in hiring Jonathon in the first place as his assistant nursery foreman?

   Do you think Mr. Nguyen took a reasonable chance on Jonathon's being able to succeed in this position?

CRITIQUE

Jonathon was well aware of what needed to be accomplished throughout the week that Mr. Nguyen was away in Seattle, but he failed in his responsibilities and did not solve basic problems that stood in the way of accomplishing his everyday tasks. Some students achieve a great deal with limited physical or intellectual ability, whereas others possess great skills and accomplish little. Some students attain only mediocre grades. Of these students, peers and teachers recognize a subset who will almost certainly make fine employees and probably in time operate their own successful horticultural nurseries. What sets these students apart from other low-GPA students is that they are diligent and motivated, striving to achieve at the very best of their ability. Jonathon was not a tremendously gifted individual, but neither was he achieving at anywhere near his potential.

Jonathon's late maturation and low motivation is all too common. The puzzle is why Mr. Nguyen hired Jonathon in the first place. He couldn't have wanted a young assistant nursery foreman whose every move he would have to direct. Ideally, Mr. Nguyen would have liked to have hired a student who excelled in the classroom and was skilled in his or her practical, hands-on knowledge of the trade. Mr. Nguyen was, however, astute and savvy enough to understand that the overall job responsibilities of the assistant nursery foreman position would not have challenged a top student for very long, nor would the salary have been attractive. He also knew that hiring a top-performing young supervisor would mean that he would have to quickly promote and give new challenges to the supervisor or risk losing him or her to a competitor. Whichever happened, Mr. Nguyen would be faced with training yet another assistant nursery foreman.

Thus Mr. Nguyen, with a relatively low-level and low-paying management position available, made the decision to offer it to a young horticulture graduate who had not been at the top of his class but worked well with his hands. In reality, Jonathon was a reasonable choice for the position and would probably have done a more than passable job if he had had more motivation.

In the end, Jonathon came to an understanding of his professional shortcomings and took responsibility for his actions. He was honest with Mr. Nguyen and had the integrity to accept the blame for his failure. Mr. Nguyen was more than fair when he allowed Jonathon to remain an additional week before going off to join the Coast Guard.

Communication: Relating to Those Around You

OBJECTIVES

After studying this chapter, you should have an understanding of:

- integrity
- trust
- ethical or professional standards
- written communication skills
- oral communication skills
- telephone communication skills
- body language
- bilingual terminology

Delegating to staff, working as a team, and transmitting a customer's requests all require that managers possess communication skills. Communication is a two-person process: one side provides information and the other side receives it. Whether managing a Fortune 500 company or a family-owned and family-operated landscape company, your communication skills directly affect its financial health and attainment of short- and long-term goals. The seasonal nature of the Green Industry in the northern regions makes communication skills much more important to have (Figure 5-1). ❧

Honesty, Integrity, and Trust

Any effort you make to communicate with your staff will not be successful if they do not respect you. Respect comes from being honest, having integrity, and being trustworthy.

What people think of your honesty has a huge impact on your ability to communicate with them. If they think you have low ethical or professional standards, or do not believe you have integrity, or do not trust you, any effort you make to communicate will be unsuccessful. Note that being trustworthy and trying to appear trustworthy are not at all the same—and your staff can

When communicating verbally,

1. initiate and maintain positive eye contact.
2. orient your body toward rather than away from the person you are speaking to.
3. do not project a defensive posture or attitude.
4. try to appear relaxed and comfortable during the conversation.
5. make sure your body language is positive.

FIGURE 5-1 Landscapers must be able to communicate their ideas clearly to clients in speech and in writing. *Courtesy of Getty Images.*

sense the difference. An ineffective manager is one whose staff consistently believes his or her stated intentions are not the true ones. Communication will always be a problem for this manager. A supervisor known to be trustworthy has achieved one of the most important milestones in his or her career.

Suppose that Robert Greene had recently moved to a small town in rural Ohio and purchased an established landscape business from John Townsend, about to retire after practicing the trade for more than 30 years. Mr. Townsend's business had a name for impeccable high quality, and the man himself was known for uncompromising honesty and integrity.

One of Robert's first landscape jobs after purchasing Mr. Townsend's small company was for Mrs. Carrie Moore, an 87-year-old widow. The planting scheme that Robert installed at Mrs. Moore's nice, not extravagant, home at the edge of town included 20 Leyland Cypress and 6 Okame flowering cherries. The specifications called for 6–7 ft tall cypress and 3.0–3.5 in. caliper cherry trees. Instead, Robert installed Leyland Cypress of 5–6 ft and cherries of 2.0–2.5 in. caliper, surmising that an 87-year-old would not notice the change and planning to use the ill-gotten profit to boost his fledgling business. Robert was correct that Mrs. Moore did not notice the inferior substitutions; however, her son, the professor of horticulture, did. Dr. Richard Moore insisted that Robert immediately replace the substitutions with the plants Mrs. Moore had paid for.

Robert Greene's lapse in professionalism will depress his business's growth for years. Although Mrs. Moore was not the type of person who would intentionally spread negative comments about Robert's work, she would obviously not recommend Robert to anyone who might inquire. In a small-town environment, this could have serious ramifications, especially if local connections were extensive, as were Mrs. Moore's. She had lived in this small town for most of her life, and her late husband had been the local pharmacist and had served two terms as mayor. Robert learned that it pays to be an honest and ethical businessperson and never to assume that a customer has no influence. A professional reputation should never be compromised. It is priceless.

Written Communication

Written communication skills are an essential part of any supervisor's movement up the organizational ladder. Supervisory staff should have the grammar, sentence structure, and overall composition skills required to clearly communicate with staff and customers. Lack of these skills inhibits advancement into mid- and upper-level managerial positions.

After basic writing skills, the most important written-communication techniques are brevity, clarity, and selectivity. Effective written communication is only as long as it needs to be. No machine has yet been invented that sifts through verbiage and extracts the essential information. Effective written communication is clearly written. If your readers cannot make sense of the e-mail, fax, memo, or report you wrote, their time and yours is wasted. Here is a five-step process to communicate clearly and effectively: (1) Organize your thoughts before you begin to write. (2) After writing your message reread it, checking for clarity of expression, for concision, and that you have respectfully conveyed the proper tone. (3) Send your message only to those who need it. You do not want to omit critical personnel but neither do you want to waste anyone's time. (4) Remember that the company name and logo and your signature are on written communications. Make sure the written communication meets overall quality standards. (5) If necessary, follow up to assure that your message's intent was accurately conveyed.

Oral Communication

Oral communication skills are as important as written communication skills. Use appropriate language at all times and insist that your staff do so also. Coarse language or foul words have no place in any business that interacts with the general public. Whether you are overseeing the installation of a large residential planting or preparing for a member–guest tournament at an exclusive country club, just one slip in your or a staff member's speech can have disastrous effects.

In oral communication, listening is as important as talking. Hearing what your supervisor as well as your staff says to you is the mark of an effective supervisor. To be a good listener, become an interactive part of the communication process. Lend an attentive ear when others are speaking. A good listener does not give the speaker one ear and listen to voice mail with the other, draft an e-mail, or search through a stack of memos.

Improve your listening skills by

1. being respectful when others voice an opposing opinion.
2. hearing what is really being said, as opposed to what you would like to hear.
3. allowing the individual to finish. Interruptions are not conducive to good communication.
4. using body language that encourages feedback and dialogue.
5. being cordial and attentive throughout even the most disruptive situations.

Telephone Skills

Telephone communication skills enable managers to speak professionally when conversing with clients in a telephone conversation. The following four techniques will help you improve telephone skills: (1) Never forget that you are talking to someone who directly supports the financial health of your organization. (2) Write down the major points you want to discuss before you make the call. This will help you focus on the primary purpose of the conversation and keep you from overlooking an important piece of information. It will also assist you in budgeting your minutes and prevent you from being repetitious. (3) Remember to speak clearly. (4) At conversation end, summarize your main points and provide the customer with your contact information.

The Finer Points of Communication

Knowing whether someone is a morning person or whether his or her biological clock ticks strongest later in the day can help you communicate. A few years ago, I worked with a manager who routinely rose at 4:30 a.m. and used the time before she left for work to balance her checkbook, wash clothes, and do other domestic tasks. She was not at her professional and productive best by the end of the workday, however. I have also worked with more than one supervisor who was barely coherent before 9:00 a.m. in the morning. Such people are virtually useless in early morning meetings, but they are often suffused with energy later in the day.

An underappreciated concept related to communication is the art of persuasion. A supervisor's ability to convince staff of a job's importance and carry it through to its completion is a worthwhile skill. No supervisor wants to have to monitor staff to see that every directive is carried out. For the most part, it is preferable for employees to take initiative in working toward the completion of a job. Having staff who require only limited guidance from their supervisor and assume responsibility means that their supervisor successfully persuaded them to buy into the project.

Body Language

Body language is an important consideration for all managers. Certain ways of standing or sitting can suggest what a person is feeling, but we have to be careful not to misinterpret others' body language. Sitting with folded arms may indicate dissatisfaction or it may represent little more than a comfortable position. Hands on hips can indicate disagreement with a specific situation or confidence and willingness to lead and take charge.

Breaking Down Barriers

According to the U.S. Census Bureau, more than 12 million second-generation Latinos have been born in the United States. By 2000 these first-generation Americans made up over a quarter of the total U.S. population of Latinos. By 2020 the number of Latinos born to immigrants will double. Corona has surpassed Heineken as the most popular imported beer in the United States, and more salsa is purchased nationwide than ketchup or mustard. Bank teller machines, medical information, and company telephone recordings are only a few examples of where Spanish–English instructions are needed (Velasquez, 2006, pp. 9, 15).

As the nation's population goes, so too goes the number of Latino workers in the Green Industry. Spanish is currently spoken in most horticultural workplaces—so much so that if supervisors do not master basic oral and written Spanish, their ability to carry out managerial responsibilities will suffer. Young supervisors need bilingual knowledge of managerial and technical terminology (e.g., *thank you, lunch, break time, shovel, wheelbarrow, mulch, mower, tree, etc.*).

Moreover, effective communication and employee safety are intimately intertwined. The safety of employees who are still learning English could be at risk if they spray chemicals or do similar routine tasks for a production nursery, golf course, or landscape maintenance company. Written warnings in the workplace in English and Spanish aid literate workers. Using symbols aids all workers, literate and illiterate alike. As with any written or oral communication, keep it simple and specific. After communicating an important message as clearly as possible, supervisors need to go further and determine whether the message was understood. Consider offering language classes for both Spanish- and English-speaking employees. At one production nursery, hour-long classes are conducted twice a week during the winter months for interested Spanish-speaking employees. The workers are paid for their class time. Supervisors who speak only English are encouraged to take introductory Spanish courses (Mulhern, 2004, pp. 59–62).

Turf, a major trade magazine that caters to industry professionals, publishes a monthly column in Spanish, "Turf en Español." The column is an excellent technical resource for Latino staff. One issue featured a training series, "Safe Lifting and Carrying Techniques," written in Spanish and English (National Institute of Occupational Safety and Health, 2005, p. C6).

Carl R. Rogers and F. J. Roethlisberger, in *Business Classics: Fifteen Key Concepts for Managerial Success*,

> propose . . . that the major barrier to mutual interpersonal communication is our very natural tendency to judge, to evaluate, to approve (or disapprove) the statement of the other person or the other group. Is there any way of solving this problem, of avoiding this barrier? . . . Real communication occurs, and this evaluative tendency is avoided[,] when we listen with understanding. What does that mean? It means . . . see[ing] the expressed idea and attitude from the other person's point of view, sens[ing] how it feels to him, to achieve his frame of reference in regard to the thing he is talking about.
>
> Stated so briefly, this may sound absurdly simple but it is not. . . . It is the most effective agent we know for altering the basic personality structure of an individual and for improving his relationships and his communications with others. (Rogers & Roethlisberger, 1952/1991, p. 45)

Rogers and Roethlisberger suggest that for communication to occur participants must not be critical or disparaging of the other's statements. Rather, each should make a special effort to appreciate what the other is communicating. Perhaps most important is that both parties be intimately engaged in the overall communication process. While one speaks, the other just as actively listens.

Listening skills are important in any personal relationship but especially so between a supervisor and his or her staff.

Encouraging effective communication in the workplace includes

1. being accepting of opinions, cultural backgrounds, and ethnicities that do not match yours.
2. not overwhelming others with your point of view.
3. striving to be clear and concise in anything you add to the conversation.
4. not forming preconceived opinions.

CONCLUSION

As we close this chapter on communication skills, let us revisit one important point: the effect staff communication has on customer service and the related financial health of the business. "Raising customer retention rates by five percentage points could increase the value of an average customer by 25–100 percent" (Reichheld, 1996, p. 33). What's more, "for every 1 percent improvement in the service climate, there's a 2 percent increase in revenue" (Goleman, Boyatzis, & McKee, 2002, p. 15).

DISCUSSION QUESTIONS

1. Why is trust important when serving customers?

   Do you agree or disagree that trust is equally important in a supervisor–staff relationship?

2. Differentiate between an individual who is trustworthy and one trying to appear trustworthy. How important is it to you whether a person is trustworthy in a professional, work-related environment?
3. What precautions would you take as a young manager if you had to work closely with a supervisor who could not be trusted?
4. Has technology advanced to the point where acceptable writing skills are no longer necessary to managers in the Green Industry?

   Why should supervisory personnel be skilled in the use of proper grammar, sentence structure, and overall composition?
5. What excuses are there for coarse or foul language in the workplace?

   What if a slipup occurs at a residence in front of the client's toddler?

   How would a respected landscape firm react if a long-established client called to complain about the coarse language used by a supervisor overseeing work at the client's residence?
6. Are telephone skills overrated in their importance to the long-term financial health of a business?

   How might a favored customer react if a young supervisor representing a long-established and respected landscape company was rude in a telephone call?
7. List three things that you would recommend most individuals do to improve their listening skills.
8. The biological clock of most individuals is weighted in favor of either the morning or evening hours. Give some examples of how this can influence the ability of supervisors to communicate effectively with their staff.
9. Is body language in today's work-related environment underrated, overrated, or neither?

   Give an example of how a supervisor could misinterpret the body language of one of his or her staff members.
10. Describe how supervisors' effectiveness in the Green Industry would be affected by their inability to speak Spanish that includes both horticultural and managerial terminology.

Robert Asbury's Communication Efforts in His First Full-time Position

Robert Asbury graduated from a major university in the Midwest with a degree in ornamental horticulture. After completing a summer internship with a firm near Cincinnati, Ohio, he assumed a property maintenance position with Evanhorst Landscape Maintenance. Evanhorst was the largest property maintenance company in the tristate region, with corporate headquarters in Chicago.

It was Evanhorst philosophy that each new supervisor would start from the ground up and grow into more responsible management positions in the organization. This philosophy came from a core belief that management personnel must have practical, hands-on experience to appreciate the work-related problems of their staff. Evanhorst had had aspiring supervisors who could not learn the task-related responsibilities of the average crew member; these supervisors were unsuccessful in future managerial roles with the organization. Robert, unlike Sarah Chen in Chapter 3's scenario, had little practical, hands-on experience and therefore started in a much lower-level position.

Robert worked out of Evanhorst's Lake County office not far from Chicago and was the newest member of a three-person crew, out of two such crews, assigned to maintain the corporate headquarters of the Berkshire Securities & Trust Corporation. Evanhorst's full-service contract gave it responsibility for the care and upkeep of all tree, shrub, and herbaceous plantings and the large acreage of turf on the site. Another prominent landscape maintenance firm initially had the contract to maintain the new corporate headquarters; however, it performed poorly and Berkshire terminated the contract after less than 4 months. Because the Berkshire complex was little more than a year old and had won numerous nationally acclaimed architectural and landscape design awards, Evanhorst assigned only its top maintenance personnel to this high-profile and lucrative account.

Both of Robert's fellow crew members were Latino, as were five of the nine other landscape and turf-related crew members at the site. Evanhorst had given Juan Cortez exclusive managerial oversight of the entire Berkshire maintenance operation. Mr. Cortez was a dedicated and highly competent project superintendent who had been employed with Evanhorst for nearly 10 years.

Robert was fully aware of the contract's importance to Evanhorst and the expectations that had been placed on him. He was highly motivated to perform at the highest level possible and was committed to working effectively with his Latino crew members. Robert enrolled in a 6-week basic-Spanish course at the local community college and then in the follow-up 8-week course. Robert memorized all managerial and technical terms in the Spanish–English/English–Spanish pocket-size handbook issued by Evanhorst to new hires. He took the initiative of requesting assistance from his Latino coworkers in pronouncing some phonetically difficult work-related terminology.

Robert learned a great deal about the practical application of horticultural techniques while employed as a crew member with Evanhorst. By the end of his first season with the firm, he had been well schooled in proper methods of corrective pruning and in deep-root fertilization of specimen trees and in general fertilizer, fungicide, and insecticide application techniques for both turf and woody ornamentals. Robert learned to drive a manual-transmission dump truck and was trained in the safe operation of a skid-steer, chain saw, and stump grinder. At Robert's request, he also spent a number of weeks employed with the turf crew at the Berkshire corporate headquarters. Through this opportunity, Robert was instructed in the operation and general maintenance of large 72 in. cut commercial-grade riding mowers as well as 42 in. wide walk-behind mowers. Robert learned the proper procedure for aerating and dethatching turf. As the season progressed, Robert identified and controlled major turf insects, diseases, and perennial weeds that had invaded the site. Mr. Cortez was very impressed with his new crew member's dedication and guided Robert in being more time efficient in his maintenance-related responsibilities.

At the end of Robert's first full season as a crew member at the Berkshire corporate complex, Mr. Cortez highly recommended him for promotion to the position of supervisor with the firm. Robert was awarded the position and was transferred at the beginning of the spring season to oversee all landscape and turf maintenance responsibilities at three dental offices as well as a midsize medical–pharmacy complex on the east side of the city. His continued diligence and hard work earned Robert a long and successful career with Evanhorst; he eventually attained the position of regional manager before he left to lead another well-respected landscape maintenance firm as their general manager–vice president of operations.

QUESTIONS FOR DISCUSSION

1. What did Robert gain by getting practical, hands-on experience as a member of a maintenance crew before being eligible for promotion to an entry-level supervisory position?
2. What did the company gain by requiring Robert to have practical, hands-on experience as a member of a maintenance crew?
3. Berkshire Securities & Trust Corporation prized its national-award-winning corporate headquarters, and Evanhorst Landscape Maintenance equally prized its role in maintaining it.

   Were you surprised that Evanhorst placed Robert on such a high-profile and important account in his first position with the firm?

   What should such an appointment mean to a new hire with ambitious managerial aspirations?
4. Although basic-Spanish courses can be valuable for a general knowledge of the language, what limitations do they have for workers in the Green Industry?
5. Robert was trained in the safe operation of a wide range of landscape and turf-related equipment. How far did such training go to better prepare Robert for ensuing supervisory positions with the firm?

6. How fortunate was Robert to have an encouraging mentor such as Mr. Cortez to guide him through his initial season with the company?
7. What additional steps could Robert have taken to further his career with Evanhorst Landscape Maintenance?

CRITIQUE

Robert was clearly a diligent and motivated first-year employee with Evanhorst Landscape Maintenance. He was aware that he needed more Spanish to work well with his crew and thus formed the goal to learn Spanish as quickly as possible.

Robert was also perceptive in seeing his need for training and adding a wide range of horticultural techniques and equipment operation to his repertoire of textbook knowledge. Most university horticulture curricula do not include anywhere near the breadth of practical, hands-on experience Robert acquired, experience required to successfully supervising staff in all areas of the Green Industry. This experience also gave him the opportunity to prove his work ethic to his new employer, which most horticultural firms want to determine early in the new hire's employment. Operating a gas-powered string trimmer for 8–10 hours in 90°F is not enjoyable nor is conveying mulch long distances in a wheelbarrow. A company tests the mental and physical capabilities of new hires who have recently graduated before promoting them.

Evanhorst knew its success with the Berkshire account rested on whether it could erase the memory of Berkshire's unhappy experience with the previous maintenance firm. Mr. Cortez was clearly valuable to the company's senior staff and trusted by them or he would not have been given the Berkshire account. Robert was fortunate that Mr. Cortez took time from his busy schedule as a skilled and responsible project superintendent to support and mentor him. Robert took his good luck and added to it by learning as much as he could from Mr. Cortez.

Before leaving this discussion, we should note that a crew member's promotion directly into a supervisor position is unusual. More often the next position a recent college graduate fills is crew leader, in which he or she has limited supervisory responsibilities and works alongside staff. The crew leader's supervisor most likely directs the priority and time frame for each task.

The ensuing position of supervisor is often the first true opportunity to make informed judgments regarding jobs or even tasks. New supervisors would have to delegate, manage their time, and prioritize responsibilities for both themselves and staff. They would have to be reasonably adept at motivating staff as well as communicating oral and written intentions. This is often the first supervisory level in which they will have the latitude to professionally interact with customers and must prove capable of representing the company's professional reputation and upholding its ethics.

However, in the Evanhorst hierarchy, the managerial position beyond crew member was that of a working supervisor. But there was another reason for Robert's quick advancement. Robert had proved himself to be a competent and reliable crew member and capable of performing at the next managerial level. Evanhorst initially had Robert oversee smaller and more centrally located accounts, where they could evaluate his progress, before moving him up the managerial ladder as he continued to grow and mature with the company.

Above supervisor, the managerial pyramid varies, depending on the size and focus of a company. Evanhorst was large, and the next managerial level was project superintendent. This position reported to an area manager who reported to a regional manager. A total of seven regional managers reported to a vice president–operations manager who coordinated directly with the company's owner. Clearly, communication skills influence a young manager's rise on the managerial ladder. Sound communication skills cannot be taken lightly.

At the Head of the Pack: How to Be a Leader

OBJECTIVES

After studying this chapter, you should have an understanding of:

- managing change
- long-term vision
- mentoring progress of staff
- problem-solving abilities
- shaping the overall culture of an organization
- organization's conduct code
- encouraging creativity
- sustained financial growth
- setting priorities
- confidence to take charge and lead
- negotiation

The ability of supervisors to lead with a vision enables an organization to compete at the highest level and provide the best possible service to clients. Good and effective leaders are the backbone of any well-run organization. ❧

What Constitutes True Leadership?

The following thoughts on leadership and its definition relate to the responsibilities of a young supervisor in the horticulture profession.

> Leadership is different from management, but not for the reasons that most people think. Leadership isn't mystical and mysterious. It has nothing to do with charisma or other exotic personality traits. It is not the province of a chosen few. Nor is leadership necessarily better than management or a replacement for it.
>
> Rather, leadership and management are two distinctive and complementary systems of action. Each has its own function and character-

istic activities. Both are necessary for success in an increasingly complex and volatile business environment.

> Management is about coping with complexity. . . . Good management brings a degree of order and consistency to key dimensions like the quality and profitability of products. . . . Leadership, by contrast, is about coping with change. Part of the reason it has become so important in recent years is that the business world has become more competitive and more volatile. . . . Major changes are more and more necessary to survive and compete effectively in this new environment. More change always involves more leadership. (Kotter, 1999, pp. 51–53)

We work and exist as professionals in an extremely seasonal and often volatile industry. The aptitude most likely to boost supervisors up the managerial ladder is that for adapting to and managing change. Change is integral to the long-term success of any organization and is a vital function of sound leadership, whether a Fortune 500 company or a small family-owned and family-operated garden center.

According to Kotter, it is important that leaders promote change. Kotter lists the following eight steps as important vehicles to be used in transforming any organization.

EIGHT STEPS TO TRANSFORMING YOUR ORGANIZATION

1. Establishing a Sense of Urgency
2. Forming a Powerful Guiding Coalition
3. Creating a Vision
4. Communicating the Vision
5. Empowering Others to Act on the Vision
6. Planning for and Creating Short-Term Wins
7. Consolidating Improvements and Producing Still More Change
8. Institutionalizing New Approaches (Kotter, 1999, p. 92)

Each principle can be related to the long-term vision of a midsize landscape company as easily as it can to redirection of the competitive edge of a multinational corporation. To incorporate these principles into day-to-day supervision, managers and supervisors must, first, delegate efficiently—that is, transfer ownership of a task to those ultimately responsible for it. Second, supervisors must create an atmosphere where taking ownership of work-related responsibilities is encouraged. Third, managers and supervisors must not neglect mentoring staff and monitoring their progress. Finally, supervisors must remain on the cutting edge of change. They must have problem-solving abilities to quickly develop strategies when coping with difficult issues. Supervisors who do not have these abilities are of little value to an organization and will be quickly replaced by more forward-thinking staff (Belasco & Stayer, 1993, p. 19).

True leaders

1. develop a long-term professional vision of where they intend to go.
2. have ethics and values.
3. are able to tackle tough obstacles head-on.
4. introduce planned and proactive change.
5. encourage creativity in their supporting staff.
6. have patience and understanding with inexperienced staff.

Principles of True Leadership

True leaders develop a vision of where they expect their role in the organization to take them and, as importantly, where they expect to take the organization. This is more difficult to implement than it sounds, and young supervisors who work with mentors will benefit greatly from their experience. True leaders have ethics and values and work toward shaping the overall culture of an organization. Horticultural companies with leaders who display these qualities are much more positive and productive places to work. Leaders must lead by example. This is true whether they are the chief executive officer of a major company or the owner of a wholesale B&B nursery. Not only should leaders have good ethics and high values but young ones should avoid associating with those who do not. It can be extremely difficult to repair a stained reputation in such a close-knit profession as the Green Industry.

To illustrate these points, let's consider Ahmed Hassan's experience at the Limca Country Club, designed and constructed in the early 1930s by a well-known golf course architect. Mike Steinberg had been the head grounds superintendent at Limca for over a dozen years and had more than 30 years of experience in all. Mr. Steinberg was an ethical, understanding, and extremely knowledgeable golf course superintendent and was a highly respected senior manager at the club.

Mr. Steinberg was convinced that his new crew member, Ahmed, had tremendous promise as a professional in the golf course industry. Ahmed had an especially keen intellect and had graduated from the university's turf management program with a 3.78 overall GPA. Even though Ahmed had achieved impressively high grades in college, Mr. Steinberg could see that Ahmed's high IQ had not been challenged in the program.

Neither was Ahmed challenged by the majority of tasks he was responsible for at Limca Country Club. Ahmed expertly operated dethatchers, aerators, greens mowers, and turf-related equipment and was equally adept at small-engine maintenance and repair. He could identify all of the major turf-related insects and diseases and understood turfgrass culture and computerized irrigation systems.

Mr. Steinberg surmised that Ahmed needed assistance in defining and focusing his career-related goals. A few months after Ahmed's hiring, Mr. Steinberg asked Ahmed if he would like to join him for lunch and discuss his professional aspirations. Ahmed had related well to Mr. Steinberg since being hired and respected his experience. He welcomed the opportunity of discussing his career goals with Mr. Steinberg. During their luncheon meeting, Ahmed confided that he did not feel professionally challenged in his current position and admitted that he lacked direction in further developing his aspirations in the turf industry. Mr. Steinberg told Ahmed that he was planning to add a third crew leader position at the beginning of next year and hinted that it might be Ahmed's if he showed he could lead by example and work to the best of his ability in his current position. Mr. Steinberg promised to try to provide Ahmed with more challenging responsibilities and suggested that they continue to strategize on Ahmed's future.

Over the next few months, Ahmed made significant progress in motivating himself and developing a focus toward his career-related goals. The following year he enrolled in three evening courses at the university, supervision and management, business accounting, and organizational theory. Ahmed progressed into the newly created crew leader position the following March and was eventually hired as an assistant superintendent in the department. After working under Mr. Steinberg's direction for an additional 4 years, Ahmed moved on to assume the director of horticulture position at Breezeway Country Club on Virginia's coast.

Ahmed, a recent graduate, benefited by discussing his career aspirations with a seasoned and experienced mentor. Although an extremely talented young man, Ahmed needed guidance in developing his future goals. Mentoring is important to future leaders in any profession. Ahmed's history of having not been challenged in college had affected how he carried out his initial responsibilities at Limca Country Club. Mr. Steinberg was wise to encourage Ahmed to look beyond his immediate role within the department and use his current responsibilities as a stepping stone to further his career. Recent graduates especially should do even the most menial tasks to the best of their ability to prove they are mature and responsible. If they fail in lower-level positions, they will have little chance of being promoted as more promising opportunities present themselves.

Characteristics of Leaders

True leaders tackle tough obstacles head-on, are success oriented, and always have a plan to help guide them in reaching their goals. True leadership is about getting the job done as efficiently and effectively as possible and encouraging others to do the same. Too many young supervisors believe that they have to take on the role of a cheerleader to motivate their crews and ultimately gain respect as a get-it-done manager. For the most part, nothing could be further from the truth. Leaders in the Green Industry are almost always high-energy individuals. Although leadership styles vary, most leaders are not extroverts or exuberant personalities. True leadership is not about flash. It is about substance.

First and foremost, true leaders introduce change but work within the organization's conduct code. Although they may be somewhat maverick in their approach to solving organizational problems, they do not circumvent company policy. You may be an extremely talented manager, one who motivates staff, is proficient at tackling large and difficult tasks, and is well known as a fiscally responsible manager; however, you must remain a dedicated team player and follow the organization's managerial guidelines. If you can't play by the rules and are constantly at odds with the goals and aspirations of the company, your future there is limited.

Second, true leaders encourage creativity. Leadership is not about inhibiting the growth and development of staff. On the contrary, it is about encouraging creativity and supporting others to reach their full potential. Highly creative employees often find it hard to work within an organizational structure. They sometimes have difficulty coordinating with their supervisors and keeping management abreast of where their creative juices are taking them. Supervisory oversight might seem inhibiting to their freedom of expression. A win–win working environment allows creativity to flow while allowing supervisors some oversight. Management of creative staff cannot take a cookie-cutter approach. *The Care and Feeding of Ideas: A Guide to Encouraging Creativity* (Adams, 1986) provides insights toward encouraging creative employees.

Third, true leaders grow and learn from past mistakes and encourage their subordinates to do the same. This does not imply that supervisors should tolerate less-than-acceptable work from their staff but that they should take age and experience into consideration when evaluating employee mistakes. Mature and experienced supervisors can sometimes be quick to chastise much-younger and less-experienced workers. Such supervisors have forgotten that they had similar problem-solving dilemmas during the formative stages of their own professional careers. True leaders always take seriously their responsibility to guide and mentor others in their managerial roles. As young supervisors grow and develop, the skills of those they supervise should mature as well. One of the legacies of a forward-looking leader is that he or she passed along experience and expertise to staff.

True leaders must

1. have and maintain confidence in their abilities.
2. have stamina and tenacity to see their goals through to completion.
3. have the vision to set long-term goals.
4. take and accept responsibility.
5. be adaptable.
6. not display an overbearing or arrogant attitude.
7. not have an overinflated ego.

Leaders come in many shapes, sizes, and personality types. All, however, are proficient in overcoming obstacles and effective in meeting the short- and long-term goals of the organization. A few other characteristics are common to leaders. Managers must contribute to the sustained financial growth of the company and be fiscally responsible. Managers at nonprofit institutions such as estates, arboretums, and botanical gardens, although not technically in the business of making a profit, also have the responsibility to be financially responsible. Leaders must be of reasonable intellect, be confident in their decision-making abilities, and have the stamina to see their vision through to completion. Arguably, tenacity in completing goals is one of the more important attributes of leaders. For the most part, being a responsible and forward-looking leader is a job that fills more than 40 hours a week. Leadership is hard work, and those willing to see each job through to completion are of incalculable value to an organization.

Leaders must be capable of setting priorities and not get sidetracked by other, less important responsibilities. Supervisors are pulled in many different directions from early in the morning until they go home at the end of the day. They must remain focused and not become a whirlwind of misguided energy. True leaders do not start 14 major projects and finish 2. Leaders often juggle numerous projects; the effective ones complete their projects with quality and consistency. Moreover, true leaders do not have to be pushed or prodded to accept responsibility.

Capable leaders do not complain, are adaptable, and display an innate confidence in their ability to take charge. Unfortunately, this last characteristic can be misunderstood by many who want to lead. The interpersonal dynamics of a group of individuals choosing a leader is likely to lead to one of three possibilities. Sometimes an individual will vie for a leadership role but be rejected by the group. In other instances, the group will see leadership qualities in one of its members but the individual will decline. The last and probably healthiest situation is when the group chooses an individual with obvious leadership qualities and confidence to take charge and lead. Capable leaders are not overbearing, arrogant, or overly assertive. Yes, it is true that good leaders are driven and are often overachievers. They are seldom comfortable to sit back and are likewise dissatisfied with placing second best. We must, however, be careful to correctly define leadership. Effective leaders do not roll over others in an effort to satisfy their own professional gain. They are likewise not driven to fulfill an already overinflated ego. Such behavior does not allow them to be a team player and does not fit the philosophy of achieving a goal for the good of the organization.

Leaders must be able to

1. always keep the interests of the organization firmly in mind.
2. use persuasion as a positive managerial tool.
3. avoid being defensive or confrontational.
4. display a positive attitude.
5. maintain an unstained reputation.
6. appreciate and respect the differing personalities of others within an organization.
7. be accountable for their decisions.

The Ability to Persuade

Persuasion is a positive component of leadership. Persuasion occurs during negotiation, a necessary and critical component of working as a team and being a successful leader. Negotiation is an important part of accomplishing goals on a day-to-day basis. In most instances, success in negotiation is directly related to the ability of the parties involved to compromise. Being defensive or confrontational is counterproductive to negotiation.

A positive and hopeful attitude is another component of successful negotiation. Nobody exudes happiness every day. Anyone who does is not in the real world. However, generally upbeat and positive individuals tend to invite cooperation. If you as a supervisor have a reputation for being positive and confident, your subordinates as well as your peers will most likely give you the support that you need.

Your reputation can help persuade others to work with you toward a unified goal. Successful leaders maintain a good reputation; that is, they are recognized as trustworthy and responsible. Having the trust and respect of others augments your ability to persuade and negotiate in times of need.

Knowing the personalities of your staff will enhance your negotiation skills. For example, suppose Mr. Rutledge, your company's general manager, has a reputation for a fiery and unpredictable temper. You have a significant problem he needs to know about. If you suspect that Mr. Rutledge is not having a good day, you know that waiting for a more opportune time will serve everyone better than the angry reaction you will probably receive if you present the problem now. However, if the problem is urgent, you have no choice about timing. Part of being a leader is being proactive, the ability to think ahead and anticipate problems before they arise and have potential solutions ready when they do.

Understanding thought processes and emotions applies to ourselves as well. We must develop a thorough and honest understanding of ourselves. We all have personal and professional deficiencies, but knowing those deficiencies means we can work to improve them. If you know you have a temper, control it rather than earning the reputation of a hothead. If time management is not your strong suit, apply yourself to becoming a better steward of the time available in your workday. If words do not always come out as fluently as you would like, take a public speaking class. In short, learn your deficiencies and try to improve them. At the very least, you will feel more confident about your overall abilities to lead and supervise staff.

CONCLUSION

Regardless of what facet of the horticulture industry you are employed in, never lose sight of the fact that leaders must accept and thrive on change. Your staff will have better attitudes and be measurably more productive if you are able to persuade them of the necessity of changes and motivate them to accept them. To do this, employees need to know the direction the organization plans to head and how management intends to get there. Communication of changes and the reasons for them aids staff in completing their responsibilities while remaining in line with the organization's long-range goals.

Supervisors must accept the level of accountability that comes with their position. It is easy to accept praise for accomplishments; accountability includes accepting responsibility for missteps. A team of talented professionals will avoid missteps. True leaders seek out and retain gifted staff to support them in their mission. True leaders not only bring out the best in their staff but also see qualities in individuals that other managers tend to overlook. They find the diamond in the rough and mold that person into a topflight employee. How many times have you seen a certain staff member perform and think, "I didn't know he had it in him." Not only did the staff member have ability but his supervisor saw it and developed it for both the employee's and the organization's long-term benefit.

The ability to bring the individual members of a team together into a cohesive whole is the mark of a leader. A staff in a constant state of disagreement gets no business done. Gifted leaders are able to get everyone to put personalities, egos, and grudges aside to see a job through to its completion. Companies search hard to find capable leaders. Those with exceptional leadership skills are hard to find and even harder to keep, but they are worth their weight in gold to the future of any organization.

DISCUSSION QUESTIONS

1. Leaders must have a vision of where they expect their role in the company to take them and, as importantly, where they expect to take the company.

   Explain how you would do this as a young supervisor.
2. What would you add to the eight basic managerial principles of leadership?
3. How do the eight basic principles of leadership relate to young supervisors in the Green Industry?
4. Why are ethics and values important in today's corporate world, including the Green Industry?
5. Give some examples of the hard decisions you as a young supervisor will have to make.
6. How does personality affect a person's ability to be a leader?
7. Highly creative employees can present managerial challenges.

   What do you envision as the toughest challenge for a supervisor of creative individuals and how would you deal with it?
8. Why is the ability to learn from one's mistakes an essential part of developing leadership abilities?
9. Why must a young supervisor always manage with the company's financial responsibilities in mind?
10. From a leadership perspective, why is it important for a leader to be persuasive and able to negotiate?
11. Why should leaders accentuate their strong points and work toward improving their professional weaknesses?

Latoya Washington, Interim Landscape Supervisor

Latoya Washington graduated with a 3.27 overall GPA in ornamental horticulture from a well-known university in the southeast and worked as a crew member at Hickory Point Country Club near Hendersonville, North Carolina. Hickory Point was a well-respected semiprivate golf course that catered to local membership as well as to visiting vacationing golfers. Although Hickory Point had never hosted a U.S. Golf Association or Ladies Professional Golf Association tournament, it had been the site of the state's Men's Senior Amateur Golf Championship 2 years ago.

Hickory Point had two separate divisions of horticulture-related employees, with both reporting to the superintendent of grounds. The Turf Division cared for all of the turf-related areas, and the Landscape Division oversaw all of the tree, shrub, and herbaceous installation and maintenance. Since early summer both divisions had been busy preparing the course for the 25th annual member–guest tournament at Hickory Point, to be held on a Saturday in October. A total of 31 pairs of golfers would vie to have their names engraved on the club's Rocha-Campbell Trophy.

Latoya Washington reported to Ken Tabata, one of two landscape supervisors at Hickory Point. The superintendent of grounds relied heavily on Mr. Tabata's guidance and expertise to assure that the herbaceous plantings would be completed in time for the member–guest tournament. Latoya and the crew she worked with were responsible for all of the landscape plantings immediately surrounding the course's clubhouse as well as annual and perennial herbaceous borders that adjoined the tennis courts, parking areas, and a recently constructed fitness facility. Seven weeks before the tournament, Mr. Tabata was injured by a hit-and-run driver while bicycling along nearby Overbrook Drive. He suffered bruises, a mild concussion, and a fractured left ankle.

Latoya had been a full-time crew member since the beginning of last year and had completed her college internship at Hickory Point the previous summer. Because Latoya's performance had been very good since her full-time hiring the previous spring, the superintendent of grounds decided to make her an interim landscape supervisor until Mr. Tabata recovered. Latoya was extremely detail-oriented and had an excellent eye for coordinating colors and textures in the landscape. Although Mr. Tabata had developed a preliminary design for the renovation of the various landscape areas, there was still much planning and installation work for Latoya to oversee. Because this year's tournament had a decidedly autumnal theme, Latoya ordered a wide array of chrysanthemums, asters, and other fall-blooming plants to complement the theme. It was the silver anniversary of the member–guest tournament, and a black-tie dinner was planned for later that evening. Latoya diligently tried to coordinate the flower colors of the mums and asters with the event's decor and decorations.

Latoya developed her daily work lists outlining the responsibilities for each of the three full-time gardeners on her landscape crew. Because Mr. Tabata's accident had interrupted his planning, many of the color combinations were not noted on the plan and needed coordination throughout many of the landscape areas adjoining the clubhouse, tennis court, and health and fitness complexes. Therefore Latoya had to further refine the preliminary designs that Mr. Tabata had begun to develop. Latoya worked closely with a nearby wholesale grower to coordinate its deliveries with the day-to-day planting of each landscape area. She constantly adjusted delivery schedules with the grower and thus avoided the need to store and care for plants before they went in the ground. In the first few weeks in her new role, Latoya spent nearly all of her time at work helping her crew members install plants. She spent many hours of her personal time planning her crew's next workday. The planning and designs allowed the superintendent of grounds to visualize Latoya's choices of plants and colors for the renovation of each landscape and review her work in the final weeks. Her color-coded computerized worksheets aided the three full-time crew members to more easily implement each design concept throughout the property.

Latoya was conscientious in encouraging each of the full-time gardeners to give their opinion each day and throughout the landscape renovation process. She held 15–20-minute strategy meetings with her crew at the beginning of each workday and also met with them after lunch to review the morning's progress and recap the work plan for that afternoon. Latoya made a personal commitment to successfully complete the project. Latoya and the crew worked until noon for several of the Saturdays leading up to the tournament, for which the club compensated them at an overtime rate. Latoya coordinated needed equipment with other supervisors in both the Turf and the Landscape divisions. Because Latoya was naturally quiet and reserved, she had to stretch herself to be more outgoing and communicative throughout the entire process. This was most helpful in the day-to-day supervision of her staff as well as in her new working relationship with the superintendent of grounds. She did her very best to keep him informed.

Hickory Point's Silver Anniversary Tournament and associated black-tie dinner was a success. The superintendent of grounds and his staff were highly praised by Hickory Point's chairman of the board and the tournament chairperson at the black-tie event.

QUESTIONS FOR DISCUSSION

1. Do you agree that working as a horticultural staff member at either a private or semiprivate country club demands an employee with the proper professional mind-set?

   What would you envision some of these professional characteristics might be?

2. How well should the superintendent of grounds at a club such as Hickory Point understand the technical aspects of playing golf to be effective in his or her position?

3. What knowledge might you gain from working at a quality-oriented club such as Hickory Point?
4. In view of the pressures associated with Hickory Point's 25th anniversary member–guest golf tournament, was the responsibility given to Latoya by the superintendent of grounds misplaced?
5. Discuss what you see as the core of Latoya's success in meeting her responsibilities as part of Hickory Point's Silver Anniversary Tournament.

   Was she professional in the way she coordinated with the superintendent of grounds as well as with managerial staff from other departments?
6. Critique Latoya's professional relationship with her crew members. Overall, was her decision making appropriate? Was she a disciplined yet amiable manager?
7. Latoya did much in her attempts to coordinate with her crew members. How might Latoya have better balanced the crew's need for supervision with her own managerial responsibilities?
8. Does Latoya have the type of personality that would allow her to grow and learn from her managerial mistakes?
9. If you had been the superintendent of grounds at Hickory Point Country Club, how would you have rated Latoya's overall success as interim landscape supervisor in preparing for the club's 25th anniversary member–guest tournament?

CRITIQUE

Latoya assumed a significant increase in responsibility after Mr. Tabata's accident, and much good can be said about her performance as interim landscape supervisor at Hickory Point Country Club. Her personal integrity and maturity along with a focus for work of the highest quality were in large part the reason for her success.

Latoya was a dependable employee, and her diligence in orchestrating the design and installation responsibilities of the landscape plantings was noticed and appreciated. Latoya showed good leadership qualities in coordinating and directing the efforts of the three full-time gardeners on her landscape crew. The color-coded computer printouts of each landscape area were a significant help to both her crew members and the superintendent of grounds in visualizing the planting concepts.

The superintendent of grounds had high expectations for Latoya, and the ball was in her court as far as coordination and communication were concerned. In a situation like this, it is essential that a young manager discuss and communicate the intended game plan with his or her supervisor. They need to be in total sync with each other in working to fulfill the goals of the organization. The last thing the superintendent of grounds wanted in a situation like this was surprises.

Latoya performed admirably in her coordination with the superintendent of grounds. When communicating with him Latoya did her very best not to catch him off guard, especially as the event drew closer and everyone's nerves became more frayed. The superintendent of grounds felt the pressure put on him by the high profile of the 25th anniversary member–guest tournament and associated black-tie dinner. Even in a normal year at Hickory Point, the job of the superintendent of grounds would not have been a low-pressure one, and for the few months leading up to the event, the expectations for him and his staff were high.

Latoya was clearly a very ethical individual and quickly showed that, although she had high expectations for her subordinates, she was a highly motivated professional in her own right. Latoya definitely led by example and was obviously willing to tackle obstacles in the way of meeting her expanded responsibilities. She did not have to be pushed or prodded to accept responsibility. Latoya was not a maverick and did not work outside the bounds of the organization. She was also not the least bit arrogant or overbearing and did not develop an overinflated ego as a result of her newfound responsibilities. Finally, Latoya had developed sound abilities toward persuading and motivating her staff.

Although Latoya's experience as interim landscape supervisor had been very successful, she could have done better in a few areas. First, her diligence in including her crew in the decision-making process took away from the time they spent installing plants, and their productivity suffered. Latoya quickly found that including staff in the planning process has to be balanced with completing day-to-day tasks. Productivity also took a hit from Latoya's giving attention to the tiniest detail while coordinating and planning plantings. To increase productivity, Latoya spent many days assisting the crew members in installing the plants. That meant she had to spend personal time on planning her crew's responsibilities for the following workday. It was obvious to the superintendent of grounds that Latoya needed to think beyond her previous role as a crew member and work toward developing a broader supervisory mind-set.

Another area of improvement for Latoya was in learning to better coordinate deliveries. Her intentions were good, but she needed to be more cognizant of the grower's need to efficiently manage his for-profit business. Fortunately, the grower had had Hickory Point as a major account for years and was a close friend of the superintendent of grounds. The grower did not know Latoya well, but he kindly overlooked numerous inconveniences in an effort to assist her through her first supervisory experience. In the end, Latoya learned a valuable lesson in supporting the operational needs of long-dedicated suppliers and vendors when making plant-related orders. Relationships with growers and other professionals who meet delivery deadlines should be cultivated.

Latoya's interim promotion threw her into a position of significant responsibility. She performed admirably in some areas and less well in others. She learned from her mistakes as a first-time manager and corrected much of the flawed decision making that had plagued her early in her appointment. The superintendent of grounds believed that Latoya had matured considerably during her appointment. When Mr. Tabata left Hickory Point the following March, Latoya was promoted to landscape supervisor, partly because of her ability to accept responsibility and high-quality work.

Delegating Authority: How to Get the Job Done

OBJECTIVES

After studying this chapter, you should have an understanding of:

- authoritative leadership
- participatory leadership
- delegative leadership
- accountability
- flexibility
- contingency plans
- problem solving

Supervisors who delegate tasks to staff draw on a complex of skills: communication, decision making, leadership, planning, motivating, and time management. ❧

Delegation and Leadership

Delegation and leadership are inextricably linked. No supervisor can delegate effectively without sound leadership skills, including the ability to practice differing styles of leadership: authoritative, participative, or delegative (Townsend & Gebhardt, 1992, pp. 25–26). The three leadership styles know no professional boundaries; they pertain to all of corporate America. Whether a crew leader for an arboriculture company or a crew leader at an exclusive country club, the leadership principles to follow remain the same.

Inexperienced managers tend to be authoritative. An authoritative leadership style seems less complex, is easier to implement, and remains attractive until the supervisor gains more experience. As supervisors begin to gain practical, real-life experience, they learn that an authoritative style inhibits staff in solving problems and squashes their creativity. Staff cannot grow and develop for the benefit of the organization, and bright and capable staff leave for positions where they are given more freedom to manage their professional responsibilities.

A participatory leadership style, on the other hand, solicits staff input in the decision-making process. This encourages them to take ownership of plans and buy into the satisfactory completion of a project. Although staff have significantly more interaction in a participative style of leadership, management still retains ultimate responsibility in the decision-making process. The most complex style of leadership and the hardest to implement is delegative. Although management still maintains responsibility for the final outcome, much of the decision making is taken on by staff. Supervisors who surround themselves with a mature and responsible staff and have a delegative leadership style will experience nearly unlimited benefits.

When choosing a leadership style, be flexible. Mesh your supervisory style with the experience and capabilities of staff and coordinate delegation efforts with your own managerial experience. When delegating important responsibilities to staff, strongly consider the long-range goals of the organization.

For young supervisors to delegate effectively, it is important for them to

1. establish a relationship of trust with their staff.
2. remain honest and ethical.
3. make sure staff has sufficient input in the decision-making process.
4. monitor staff progress.
5. prioritize effectively.
6. be sure that each staff member is treated equitably.
7. be sure that each staff member understands his or her responsibilities.

Delegation and Communication

Young supervisors must understand that delegation is only as effective as the communication skills used and that it should always involve two-way communication. Attempts at delegation will be wasted if your crew does not understand what is expected and within what time frame or if your crew does not inform you of impediments. Employees respect supervisors who are open to their suggestions. Honoring the opinions of staff and fostering a healthy professional relationship in working toward a common objective is the hallmark of a seasoned manager. Delegation and trust are close relatives. Managers get more done when they have earned the trust and respect of their staff. Young supervisors must be consistent, adaptable, and continually diligent in working toward improving the team.

Fine-tuning Delegation

New supervisors often believe they have to oversee every detail. Often, they must train themselves to overcome the urge to do everything themselves. At the same time, supervisors must remain aware of major issues and stay abreast of the task at hand. Understand that not every job your staff completes will be done to perfection, and sometimes staff will require correction for sloppy or inconsistent efforts.

However, not every job requires an A+ level of effort, and it is impossible to take every task to the highest level of perfection. This does not mean that an important task should be given a cursory amount of attention, or that a C-level job should be given A+ priority. Some jobs demand only a minimal level of effort. Effective managers know how much effort each task deserves.

Delegating Efficiently

Supervisors who do not delegate spend too much time directing low-priority jobs and are much less efficient in their own responsibilities. They work long hours, arriving at the office early and remaining late. But a quantitative measure of their productivity would make it clear that their value to the organization does not correspond to the amount of time they devote to it. Well-organized managers are more likely to delegate and tend to be better able to overcome the obstacles that often intrude in a workday. They are much more adept at prioritizing their professional responsibilities and adhering to the long-term goals of the organization.

Coordinating the right team with the task-at-hand does much to ensure its successful completion. Is Amy too analytic and detail-oriented to oversee this creative project? Would she not be able to visualize the entire scope of the job? Does John have the personal drive and initiative to keep pace with others to complete this other, long-term project? Will he respect its deadlines and work within the budget? For this other, complex project that requires staff to pull together, is Robert enough of a team player? Will Sara's intensity and personal motivation allow her to share responsibilities with others and not take control of it? Will Jan strive to complete the job by herself?

For subordinates to be truly responsible for assigned tasks, they must fully understand expectations. Give clear, concise, and straightforward guidelines, and define a time frame. Your staff needs to know whether a job is to be completed today, by the end of the day on Friday, or by the end of next month. Apprise them of any budgetary restrictions associated with the task. Make it clear which crew members are ultimately responsible for the completion of the job. Situations can become confused when responsibilities overlap and staff begin to trespass into each other's area of expertise. If multiple staff members are to share responsibility for a task, you need to establish clear lines of accountability (Figure 7-1). Be sure everyone knows the ultimate goal they are aiming for. Ensure that each staff member knows how he or she fits into the overall plan and that no significant questions remain unanswered.

After all assignments have been made, all responsibilities have been delegated, and all goals have been established, determine the level of communication that needs to occur throughout the life of the project. Updates and progress reports are important for tracking the scheduled versus real-life progress of the project. Appropriately delegating tasks to staff and entrusting them with increased accountability will make them much more efficient in their work. In addition, staff will have a higher level of confidence, better morale, and more enthusiasm in completing their responsibilities.

When delegating, it is important that staff understand

1. the level of accountability that is expected.
2. that it is imperative to respect timelines and budgets.
3. that equipment must often be shared and its use coordinated.
4. the importance of a contingency plan.
5. how to problem solve their way through difficult situations.
6. the value of alternative options as suggested by coworkers.
7. the benefits of working together as a team.

Productivity and Coordination

Jobs should be completed on time and under budget, whether in the private, public, or nonprofit (arboretums, botanical gardens, etc.) segments of

When delegating, young supervisors must understand the importance of

1. thanking their staff for completing assigned tasks, especially if they put in extra effort.
2. promoting a quality work environment.
3. accepting both negative and positive feedback from staff.
4. evaluating how well each major task was completed.
5. accepting and learning from their mistakes.
6. allowing capable staff to become part of the decision-making process.
7. mentoring inexperienced staff toward the completion of an important goal.

DATE	**MAINTENANCE ACTIVITIES**
8/23	Joe B. 8hr. sodding football field - 4hr fertilizing fields
"	Sally K. 2hr. sodding football field- 6hr mowing fields
"	Al. M. 2hr. weeding flower beds- 6hr fertilizing lawns
"	Heather S. 2hr. weeding flower-beds- 6hr mowing lawns
8/24	Joe B. 1½ hr. irrigating fields - 3hr seeding - 3½hr core cultivating
"	Sally K. 4½hr. seeding - 3½hr mowing lawns
"	Al M. 5hr. fertilizing lawns - 3hr trimming around buildings
"	Heather S. 8hr. mowing lawns

FIGURE 7-1 A sample daily time-use record.

the Green Industry. Doing so might require coordination with other teams. Supervisors each have their own method of carrying out responsibilities, so make sure that you and your staff work together as a team and coordinate responsibly with other supervisors and their crews. Equipment such as front-end loaders, pickup trucks, dump trucks, and chain saws may need to be shared and coordinated between staff. Coordinate with managers if multiple tasks need to be completed in a certain sequence. This is especially true where two, three, or more crews are involved in differing areas of responsibility. Even if you as a supervisor are responsible for only a small component of the project, you must make every effort to understand and work toward the larger whole. For example, suppose the landscape company you work for has been hired to install a complete landscape package at a new, upscale corporate office complex. The landscape job would not get done on time and under budget if the petunias, geraniums, and other annual flowers were installed before the area's rough grade had been established. Likewise if the hardwood mulch were put down before the underground irrigation system had been installed or shrubs and herbaceous ground covers were planted before the large-caliper shade trees.

Accountability and Managerial Support

True accountability relies on having enough flexibility to do the job. How much is enough depends on the quality and experience of your staff and the complexity of the task. Subordinates must be capable of the responsibilities entrusted to them. To help ensure that staff carry through on their responsibilities, do not manage too stringently. There is a definite balance between managing too loosely and grasping the supervisory reins with a tightly clenched fist. Honest and open communication encourages your subordinates to keep you abreast of task progress and alert you when things go awry. For the most part, mature and seasoned staff seldom require much direction in seeing their tasks through to completion. With experienced staff, young supervisors should be more concerned with communicating the task and making sure staff have the resources to satisfactorily complete it.

Think ahead and develop contingency plans in case problems arise. If problems do arise, quickly get to their root. Lackadaisical problem solving costs a company time and money, possibly eating away at the profit margin of an otherwise lucrative landscape or turf installation contract. Encourage staff to generate new ideas and problem solve on their own. It is not uncommon for mature and seasoned staff to have significantly more technical expertise than the young supervisor they report to. What the recent college graduate should be providing in such situations is not so much the technical expertise but rather the managerial support that staff needs to do the job.

Effective supervisors seldom get bogged down in the completion of a single task, but rather concentrate on the coordinating the overall project. This final project review should include staff feedback to improve the overall success of future projects. Evaluating and reviewing the work of staff is not pointing fingers or laying blame. It is rather an opportunity to encourage staff to grow and mature in their positions and for improving the timeliness, quality, and overall efficiency of their work.

Important components of delegation include

1. encouraging staff to work through to the completion of each task.
2. not letting staff with negative attitudes influence other members of the team.
3. mentoring younger and less-experienced staff.
4. encouraging staff to solve problems according to their professional capabilities.
5. encouraging honest feedback from staff.
6. anticipating problems before they arise.
7. being fair and consistent when dealing with staff.

CONCLUSION

We will follow Jason, a young supervisor charged with the planting of a landscape border, in his duties to give us a better understanding of delegation. Jason instructed his crew leader, Alex, to plant a border of 50 butterfly bushes, *Buddleia davidii,* along the western boundary of a property owned by Dr. and Mrs. Albert Smith. The Smiths enjoy visiting their small 37-acre farm just outside Winchester, Virginia, on weekends throughout much of the year, and they particularly look forward to extended visits in June, July, and August. Mrs. Smith has a special affection for the upright bloom clusters of the 'Pink Delight' butterfly bush, with its showy, summer-blooming flowers.

Alex was not normally assigned to Jason, but he was under Jason's supervision at the Smith residence for this day. Jason, when detailing the day's responsibilities, was well aware that Alex was not one of the company's most responsible crew leaders. Jason had planned to check on Alex's work later that morning, but brush fire after brush fire demanded Jason's attention throughout the entire workday. Instead, Jason decided to stop by the Smith residence that evening as it was on his way home.

As he got in his pickup truck to leave for home, he listened to a voice message from his friend Rusty inviting him to an impromptu barbecue. Jason happily returned the call and assured Rusty he would be there in an hour. Somehow, it no longer seemed necessary to check on Alex's work.

As you have probably guessed, Alex and his crew left the job site in an unacceptable condition. Two small but very noticeable piles of debris were left along the walkway leading to the Smith's home. A stack of 20–25 black-plastic nursery containers were left propped against the Smith's horse barn. If that were not enough, Alex had backed the company truck too far off the Smith's driveway and damaged the dry-laid fieldstone wall surrounding the swimming pool and patio. Extremely disappointed with the quality of the crew's work, Mrs. Smith left a voice mail for the director of landscape services late that evening about the condition of her landscape. The director of landscape services knew that Jason had planned on checking the progress of the border planting during the day. He obviously had not.

Jason was well aware that planting 50 butterfly bushes was not a large or complicated job. He also knew that Dr. and Mrs. Smith were longtime and valued customers of the landscape company. And he knew that Alex was not the most reliable crew leader. Jason had no other choice than to send Alex to do the job. He could have compensated for the unreliable Alex by making time to check on Alex's work. As often happened to him, Jason had been caught up in the many small details that develop in the course of a day. Jason was indeed busy from early morning to the end of the day, but he was being busy and not productive. He should have divorced himself from the many brush fires that sidetracked him to get out to the Smith's residence by the end of the day. Alex himself was at fault for much of what went wrong at the Smith's residence, and he lost his job the following day, having been irresponsible one too many times in his brief employment with the company. Jason himself was irresponsible in not making sure that Alex's work was completed to the highest level of satisfaction. An important part of delegation involves following up on staff. Such supervision can prevent an unexpected situation from becoming a misfortune.

Jason should have remembered the difference between performing the minimal obligation to his employer and serving the company's best interest. Sometimes managers need to look beyond the list of duties that may be outlined in their job description to make mature and responsible decisions. If Jason had checked Alex's work that evening, he could have easily put the nursery pots in his pickup truck and removed the two small but prominent piles of debris. He could have also apologized to Mrs. Smith for the poor performance of his staff—before she called the director of landscape services. Instead of a complaint, the director might have heard how Jason had made an effort to correct the situation and had personally picked up the pots and debris. Jason did not have the authority to directly address the company's rebuilding of the fieldstone wall, but he could have assured Mr. and Mrs. Smith that the director of landscape services would have it repaired quickly. Although the mess was Alex's, Jason was ultimately responsible and should have been more diligent in overseeing and directing the responsibilities of his staff.

DISCUSSION QUESTIONS

1. Describe how communication is a vital component of delegating to staff.
2. How are delegation and staff teamwork related?
3. Explain how delegation, trust, and respect must coexist for a positive supervisor–staff relationship to exist.
4. Explain why it is important to empower your staff and encourage them to problem solve.

   Why should supervisors stay abreast of important tasks and monitor staff progress?
5. Why should some jobs receive only a minimal level of effort toward their satisfactory completion?
6. Being busy does not necessarily mean that you are being productive in your work-related responsibilities. How does this relate to organizational skills and the ability to delegate and manage staff?
7. How are time management, communication, and delegation linked?
8. How are responsibility and accountability related to a supervisor's ability to delegate?

   Why does there need to be a balance between giving subordinates responsibility and managing too tightly?
9. Why might a long-established manager not understand that young supervisors should be given sufficient time to develop their problem-solving abilities?
10. How would you have supervised Alex, knowing his deficiencies as a crew leader?
11. How could Jason have better managed his time and prevented himself from getting caught up in the many brush fires that compromised his effectiveness as a supervisor?

Liz, Tamika, and Jessica's Development of Laurel Gardens

Liz Henderson, Tamika Gardner, and Jessica Roberts were roommates at a large and respected university in the southeast. All majored in ornamental horticulture or landscape architecture with the hope of developing a long-term future in retail sales. Each of these young women was an excellent student and was dedicated, motivated, and responsible in meeting career goals.

Liz, Tamika, and Jessica decided to start a combination retail garden center and landscape design and installation business called Laurel Gardens. Each had interned during the summer of their junior year at college at a respected garden center to gain hands-on experience. Liz had interned at the Hollygreen Home & Garden Center in Westminster, Massachusetts; Tamika at Rossetti Gardens just outside Newport, Rhode Island; and Jessica at Wellington Nursery & Gardens, not far from her home in Hyde Park, New York. Hollygreen was a large regional company comprising 23 home and garden centers in the mid-Atlantic states; Rossetti Gardens and Wellington Nursery & Gardens were privately owned and operated businesses.

The young women had planned their internships so that they could collectively benefit from experiencing the management styles of a large, corporate-based organization and two family-owned and family-operated businesses. During their senior year at college, Liz, Tamika, and Jessica researched many potential-competitor garden centers and landscape companies within a 30-mile radius of Charlottesville, Virginia. They reviewed Charlottesville's economic status and evaluated its growth trends over the previous decade. It was clear that Charlottesville and its neighboring communities housed an acceptably high percentage of middle- and upper-middle-class professionals in two-income households. Liz, Tamika, and Jessica had visited a commercial real estate firm in Charlottesville during the spring break of their senior year at college to evaluate possible locations for their business. They also met with loan officers at three of the local banks and investigated lending opportunities.

With accumulated savings from their summer jobs, some much-needed financial assistance from each of their parents, and a large loan the young women signed a lease on a desirable parcel of commercial property. The site had most recently housed a well-established privately owned farmer's market. The previous owner had retired after operating a successful fruit, vegetable, and bedding-plant operation for 27 years. The corner lot housed a well-kept two-story brick building suitable for both retail and office space. The site included a gravel parking lot of adequate size and ample room for future lathe structures, polyhouses, and an outdoor sales yard.

Because Liz had a minor in business management with an emphasis in accounting she would oversee financial operations. Tamika had an outgoing personality and was especially gifted in customer service. She had acquired sound experience in retail sales and had excellent knowledge in the identification and culture of woody and herbaceous plants. Tamika would oversee the day-to-day operation of the garden center and adjoining outdoor sales yard. Jessica would use one of the upstairs rooms as a design studio and sell upper-end landscape designs to a select group of specialty clientele. Jessica had an excellent knowledge of native plants and a skill for developing unique designs. She also planned to supervise the three-person landscape crew that would install her landscapes.

Liz, Tamika, and Jessica did not graduate until late May and were unable to capitalize on the busy spring season. They worked long hours throughout the hot, dry summer months to get the business up and running by early autumn. Although they had grossly underestimated the amount of preparation and planning needed to initiate a new business venture, they were able to open the doors of an attractive and well-appointed garden shop on a Saturday in September. The adjoining lathe structure and sales yard contained a small but well-displayed selection of ferns, ground covers, and flowering herbaceous ornamentals. Tamika had secured a complement of B&B and containerized woody trees and shrubs. She was also able to purchase an excellent selection of quality herbaceous perennials from a local grower.

Growing Pains

That 1st year was an extremely difficult and frustrating one for the three new entrepreneurs. They all agreed that they would do things differently if they could have the chance to start over. Locating and managing competent staff turned out to be a major headache. They found enough seasonal employees to work in the garden center and on landscape crews, but finding responsible and motivated ones was another matter. Throughout the fall planting season, some of the hourly staff arrived late to work or did not show up at all. Liz, Tamika, and Jessica spent an excessive amount of time on menial tasks that they should have been able to delegate. Their attention to demanding minor details kept each of them working long hours in an effort to maintain the financial health of the business.

Liz, Tamika, and Jessica had wrongly assumed their staff would give company property the same care and respect they did. Their first autumn season, they acquired a small pickup truck for Jessica to use in meeting with her clients and a full-size, heavy-duty pickup truck and accompanying flat-bed trailer for use by their three-person installation crew. It soon became obvious that their crew leader gave little oversight to the two laborers under him. Their pickup showed premature wear. Jessica was forced to take valuable time away from designing and selling her jobs. Overseeing each landscape planting and assuring the designs were installed with the promised pride and attention to detail that the business was to be known for became her overriding concern.

Jessica found designing as well as overseeing installations to be extremely time-consuming and difficult. To assure that their reputation did not suffer, the new owners decided to eliminate the three-person planting crew and contract with a small but quality-oriented local landscape firm to install Jessica's designs. Jessica, freed from labor-related issues, was able to focus on her designs again.

Tamika found she had underestimated the patience a garden center owner must have. Although Tamika had prided herself on being able to serve the most difficult clientele, she was surprised at how emotionally taxing such customers can be when dealing with them day after day. Tamika also had trouble finding growers and hardscape vendors who could be trusted to meet their agreed-on responsibilities. She made and terminated agreements with two perennial suppliers before she found a grower who could dependably provide high-quality plants at a reasonable wholesale price.

Finally, even with Liz's academic background in accounting, she had underestimated how erratic cash flow could be during a company's first few months of operation. Although the three entrepreneurs worked hard and eventually made Laurel Gardens a success, it took much more time and effort than any of them had ever imagined it would.

QUESTIONS FOR DISCUSSION

1. Was the practical horticultural and business knowledge of Liz, Tamika, and Jessica sufficient for them to start Laurel Gardens?
2. What did the internship experiences of Liz, Tamika, and Jessica add to their future operation of Laurel Gardens?

   What would Liz, Tamika, and Jessica have lost if they had not interned at different garden centers?
3. What are the advantages and disadvantages of close friends operating a business venture together?
4. Were Liz, Tamika, and Jessica diligent in fully researching the full range of variables associated with the initiation of their small business venture?
5. Were Liz, Tamika, and Jessica foolish to take out such a large loan and establish a full-service garden center and landscape business so soon after graduating from college?
6. What did the three omit from their plans that would have warned them of the hard work and long hours necessary to establish Laurel Gardens?

   How could Jessica have avoided the labor-related issues she had to deal with in the initial stages of the business?

7. How could Tamika have avoided the problems she had in locating a reliable vendor to supply quality perennials at a reasonable wholesale price?
8. How could Liz have avoided the frustrations she had with the cash flow issues of a new business venture?
9. What advantages would Liz, Tamika, and Jessica have seen if they had purchased a good-quality, preowned pickup truck for their landscape crew instead of the new vehicle?
10. How should Liz, Tamika, and Jessica have amended their business strategy in the opening of Laurel Gardens? Be very detailed in your answer.

CRITIQUE

The benefits of being your own boss always comes with a price. Small, privately owned businesses face many pitfalls, and an alarmingly high percentage of new businesses fail in their first 3–5 years of operation. Although the start-up of any company has risks, Green Industry businesses seem to be particularly vulnerable. Because of the high cost of equipment—front-end loaders, heavy-duty dump trucks, commercial-grade mowers—garden center and landscape businesses are expensive to start. They are extremely weather dependent, particularly in northern states, and labor dependent and highly competitive. Although starting your own business can be a wonderful and personally rewarding venture, it can also be a nightmare for the unprepared, whether horticulturally or financially.

That's not to say that a business couldn't have low start-up costs and eventually succeed even when the owner has little business experience. With diligence and hard work, a fledgling business that purchased equipment to cut small to midsize residential and commercial properties and had a reliable pickup truck and trailer could develop into a full-service turf management company with a reasonably large complement of both full-time and seasonal employees. However, except in rare instances, having saved a sizable amount of capital and gained experience before you start your own company will greatly increase the chances that your new business will survive and prosper.

Let's analyze some of the decisions that Liz, Tamika, and Jessica made as new business owners to see if we can learn from their mistakes. First, they should be commended for their diligence and the time they spent researching their business. Businesses started by many recent graduates have two main risks: lack of experience and lack of money. Although students might believe they have gained reasonable horticultural knowledge, academic study or coordination of side jobs during college cannot replace the managerial and business-related experience necessary to the full-time operation of a Green Industry company. For example, students running side jobs while in college frequently do not provide medical benefits to full-time employees. Their businesses may not be insured, and they have not yet begun paying off school loans or loans on buildings and equipment. They may have little experience overseeing a sizable full-time or seasonal labor force or orchestrating installations costing tens of thousands of dollars. Some students might have the maturity, experience, and even financial backing to start their own enterprises directly after leaving school, but the large majority will significantly benefit by first gaining full-time experience in the trade.

It takes significant financial resources to start a Green Industry business and more to run it in its first few years of operation. Liz, Tamika, and Jessica were extremely ambitious in their plans and were not completely prepared for the many pitfalls common to a new business. Small finances and big plans coupled with unexpected expenses add up to hard times. They should have avoided taking out such a large loan as they had not yet had the opportunity to save significant capital before starting such an ambitious enterprise. Sound financial management is fundamental to any small business. If a recent graduate borrows heavily to start a new business and that business fails, how will the outstanding debts be repaid? In addition to Liz's academic background in business and finance, Tamika and Jessica could have taken similar course work.

Most new business owners have more success if they do not try to start their own business immediately after graduating from college. Anyone interested in starting a small horticulture business should have some background in small-business and financial management. Courses in business law, accounting, supervision and management, organizational theory, small-business management, microeconomics, and business administration are minimal requirements for anyone considering starting a garden center, landscape operation, or nursery business.

It is seldom wise for new graduates to purchase new equipment. If you have little business experience and a relatively small amount of capital to invest, it makes little sense to purchase a new heavy-duty pickup truck, 60-in. commercial mower, or front-end loader. A long-established business could have sound reasons for buying large and expensive new equipment, but remember that this is work equipment and bound to get dings, scrapes, and scratches. Buying a well-cared-for, preowned truck from an honest and reputable dealer who guarantees vehicles can save you thousands of dollars. Moreover, most clients don't really care if you drive up in a new pickup truck to install their landscape designs. They are much more concerned with your reputation and the quality of your work.

The mental, emotional, and physical preparation required to start a new business is immense. On top of that is the difficulty of locating first-rate employees. Careless and misdirected employees can make a new owner's job infinitely more difficult. Jessica's decision to subcontract the installation of her planting designs with a reputable landscape firm may or may not have been a viable long-term solution to the problem. It did, however, allow her to handle the immediate situation.

Liz, Tamika, and Jessica were bright and talented entrepreneurs, but their ambitions nearly overwhelmed their financial resources and capacity for dedicating long hours to it. New graduates need to carefully think through their business plans, get advice from a mentor, and ask themselves if they are being too hasty in achieving their long-term goals.

Understanding the Big Picture: Managing the Fiscal Health of Your Business

OBJECTIVES

After studying this chapter, you should have an understanding of:

- total average revenue
- net revenue
- average operating cost
- average net profit
- customer service
- customer base
- budgeting process
- rolling budget
- incremental budget
- zero-based budget
- balance sheet
- income statement
- cash flow statement
- business plan

Most business owners take too limited a view of managing the fiscal health of their companies. Although the information in balance sheets, income statements, and cash flow statements is critical to the future of any organization, there is much more to charting the financial course of a business. The long-term success of a business relies on coordinating three components: First, business owners must be attuned to the niche their company occupies within the trade and be in sync with current and anticipated trends within the industry. Second, business owners must be receptive to customer needs. They must know their customers and fully appreciate how important their customers are to the bottom line of the company. Third, business owners must understand the information presented in financial statements. They must be able to use this information in long-term decision making. ❧

Landscape Industry Finances

Business owners must maintain a solid understanding of their niche in the industry and have an intimate appreciation of how their business relates to the competition. Moreover, business owners cannot manage the financial future of their company if they do not accurately perceive where the industry is heading in the long term. A 2005 survey of the landscape industry found that the total average revenue per company increased by more than $60,000, net revenue was up 1.8 percent, and service prices increased nearly 12 percent. Average annual revenue was nearly $800,000, and the average net profit increased 11 percent. The average age of companies in the survey was 13.5 years, although 30 percent had been in business for less than 5 years. Respondents employed an average of 6.5 year-round employees and slightly more than 10 seasonal workers. Most of these figures are encouraging, but owners of the surveyed companies had significant concerns about the future of the industry.

The survey ranked the top five concerns the business owners had for 2006:

1. Fuel prices
2. Worker compensation costs
3. Lowball competitors
4. Health insurance and overworked or overstressed employees
5. Inflation and rising interest rates

One of the most pertinent facts gleaned from this survey was that landscape contractors had increased their use of subcontractors, spending an astounding 72 percent more on them, or nearly twice the previous year's amount. The reason for the increase is mostly related to rising health insurance costs, but some increase might be due to lowball competitors and overworked or overstressed employees. Lowball competitors are a concern throughout the Green Industry, not just in the landscape trade (Wisniewski, 2005).

Along the same lines, it is not uncommon for new arboriculture business ventures to close their doors after just a few years after opening. After a few years of operation, new owners have exhausted work opportunities provided by friends and relatives and the once-new trucks, chippers, and chain saws show wear. Expensive repair bills and lack of work mean the owners can no longer meet their loan payments and still profitably operate the business, explaining the short life span of some arboriculture companies.

Customer Service and Fiscal Health

Customer service and the fiscal health of a company are inextricably linked. Good customer relationships and a dedicated core of customers are directly related to a business's long-term fiscal health. Determine your niche and identify which clients you can best serve. Establish a competitive advantage in serving those clients, and provide them with your full attention. Incorporate this customer-oriented philosophy in the day-to-day management of your business (Whiteley & Hessan, 1996).

To understand your customers you need to know their demographics. Know the age, marital status, education, mobility, and socioeconomic levels of the types of customers you serve. Gain an awareness of how dedicated customers are to the business and whether they are willing to pay a premium price for quality services or would rather shop for the lowest price. To set long-term goals, business owners must be aware of how their customer base is changing and the changes' effect on the company's fiscal health. Determine whether customers come to you as a result of word-of-mouth recommendations or paid advertising. If paid advertising has drawn in most customers, which media have been most successful in reaching out to clients? Stay abreast of trends in advertising and create new advertising strategies if needed (Wayland & Cole, 1997, p. 103).

The Green Industry is composed of some large corporations but many are small to midsize businesses. The biggest moneymaker of them all amassed total revenues exceeding $1.4 billion during 2004; the 100th-ranked firm, $11 million (Wisniewski, 2005).

The Budget

Budgets have a large role in planning the financial future of a business because they allow managers to stay abreast of where the business is today and know where it is heading tomorrow (Figure 8-1). A budget that is carefully planned, accurate, and up to date reflects the long-term goals of the company.

> The budgeting process forces you to estimate how many of each product or service you will produce and sell, the cost of those items, the pace at which receivables will be collected, general expenses, and taxes. These figures provide a forecast for the months or year ahead. A good budget helps you assess whether or not the business will have adequate financial resources to stay the course. (Harvard Business Essentials [HBS], 2004, p. 218)

Moreover, budgets assist managers in keeping their finger on the financial pulse of the business. If the fiscal health of a company remains on track, management has little need to amend its goals. If fiscal health runs askew, however, management can initiate contingency plans. For example, Blue Mountain Landscape & Design had budgeted for acquiring a new dump truck in early June of the upcoming year. An unseasonably wet spring prevented the company from completing by mid-May the scheduled plant and hardscape installations. The number of clients serviced therefore remained far below the company's forecasted goals and cash flow lagged considerably behind expectations. Reviewing the financial status of the company, the owner decided to revise plans and not purchase the dump truck.

CATEGORY	2007 BUDGET
Salaries and Wages	$771,312
Payroll Taxes	$115,697
Uniforms	$3,563
Fuel/Oil/Gas	$23,665
Paper Goods/Cleaning Fluids	$1,000
Course/Office Supplies	$14,231
Equipment Maintenance	$35,988
Electric	$25,011
Permits	$2,630
Equipment Repairs	$27,000
Telephone	$2,500
Batteries & Tires	$2,000
Education	$9,420
Tree Maintenance	$30,000
Fertilizer	$34,106
Fungicide	$66,087
Herbicide	$17,852
Insecticide	$22,642
Seed	$52,804
Topdressing Sand	$19,045
Bunker Sand	$2,500
Roads & Paths	$17,366
Landscaping	$4,850
Irrigation Repairs	$9,000
Irrigation Maint.	$2,725
Drainage	$825
Superintendent Expenses	$3,410
Small Tool Replacement	$6,000
USGA and Weather Service	$3,305
Compost Management	$2,250
Trash Pickup	$3,238
Soil Tests	$3,668
Miscellaneous	$4,850
Total: Golf Course	**$1,340,540**

FIGURE 8-1 Proposed annual budget.

The type of budget to use depends on the day-to-day operation of the business. Most companies operate on a yearly financial cycle, but a longer or shorter time frame may better suit a company's needs. Most firms' budgets cover a fixed period. A rolling budget is favored by companies that demand more flexibility in evaluating the fiscal health of their operation. Although rolling budgets can be constantly amended, they tend to be time-consuming to manage. The two types of budgets primarily used are incremental budgets and zero-based budgets. An incremental budget is guided by the budget that preceded it. Information from the previous budget, coupled with financial projections for the upcoming business cycle, helps develop the next budget. Zero-based budgets have the advantage that each successive budget is developed anew and is a separate financial entity unto itself.

No one is born knowing how to produce a budget, and young supervisors should seek guidance from a seasoned manager. When creating a budget, dedicate sufficient time and energy to the process to make the budget a quality document. Use only accurate and up-to-date information. A budget should conservatively reflect the long-term fiscal goals of a business; someone unsure of company goals will produce a budget that is incomplete at best. Finally, when making budgetary requests, managers should be realistic and remain within the organization's financial framework (HBS, 2004, pp. 238–240).

Financial Statements

Interpreting financial statements—balance sheets, income statements, and cash flow statements—is a necessary skill for any young supervisor.

The Balance Sheet

A balance sheet provides an overview of the fiscal health of a business. Balance sheets summarize the financial condition of a company and are frequently developed at predetermined times throughout the fiscal year, often on a monthly or quarterly basis. Some are published at the end of the fiscal year. A balance sheet adds up the assets and subtracts the liabilities of a company. The final result of this equation is defined as owner's equity. Capital is the total current investment in a business. Assets include land, buildings, equipment (less its accumulated depreciation), supplies, and other tangibles. It also includes profits that have been retained in the business. Cash, stocks, and similar investments and accounts receivable, or what creditors owe the company, are also assets. For example, the accounts receivable of a B&B nursery would include the wholesale value of pin oaks, *Quercus palustris*, recently purchased by a landscape firm.

A company's balance sheet

- provides a financial picture of the company at any given point in time.
- provides an accurate measure of the company's financial stability.
- lists assets, or items of value, owned by the company.
- lists liabilities, or debts, of the company.

The balance sheet lists financial liabilities the company has incurred. For example, if a wholesale nursery recently purchased a large quantity of wire baskets to use in digging B&B trees, the unpaid invoice for the wire baskets would be included as a liability. A company's liabilities may include accrued expenses, income tax payables, and long-term debt. The balance sheet lists, first, current liabilities, or those that would normally be paid off within 1 year of the balance sheet's date. Current liabilities include a range of categories, from accounts payable and accrued salaries to short-term debts. Long-term liabilities, those debts not due for at least a year, are listed next and include, for example, a 20-year mortgage on the main office building or a 15-year loan towards the purchase of additional land to increase the size of a wholesale nursery. However, the current portion of a long-term debt, such as mortgage payments, would be included as a current liability. Notes payable are debts owed on money borrowed.

Assets are listed on a balance sheet according to their liquidity, or ability to be converted into cash. For example, those wire baskets whose invoice is a liability would themselves be an asset. Other liquid assets include marketable securities in the form of stocks or accounts receivable. The balance sheet next lists assets more difficult to turn into cash, such as land, buildings, and equipment. These are normally designated as fixed assets. Many fixed assets, such as equipment, tend to lose value over time, or depreciate. Subtracting depreciation provides a more realistic picture of fixed assets' worth, or current book value.

The Income Statement

An income statement depicts the cumulative financial health of a business over an extended period. Another name for an income statement is a profit-and-loss statement. The income statement shows the relative profitability of a company. The income statement begins by noting the revenues the company generated over a specified period. For example, a commercial B&B nursery receives revenue from the sale of its trees and shrubs to a landscape company or garden center. The landscape company or garden center resells these same trees and shrubs to their retail clients at a profit. Companies can have additional sources of revenue aside from the products or services that constitute the mainstay of the business. As an example, a wholesale nursery may receive dividends from stock market investments or monthly payments from renting out a storage facility. A nursery might lease a portion of its acreage to a neighboring farmer. An income statement also shows significant expenses, which indicate costs associated with day-to-day operation of the business. A major expense of a wholesale nursery would be its cost of goods sold. For many industries the cost of goods sold is the cost to purchase raw materials or the cost of growing or fabricating raw materials into a finished product. As an example, the cost of goods sold for a wholesale B&B nursery might include purchase of lining-out stock.

Another large expense for most companies is the cost to operate. Operating expenses are indirectly associated with the cost of growing or manufacturing a product. Examples of operating expenses include administrative costs, advertising expense, utilities, and staff salaries. After all of the incurred expenses have been subtracted from the revenues, the result is the firm's net income. If expenses exceed revenues for the corresponding period, the company failed to realize a profit and is said to have incurred a net loss.

The income statement

1. has a monthly, quarterly, or yearly period.
2. details the net profit or net loss for the period indicated.
3. lists interest income, or revenue from money invested.
4. lists cost of goods sold, or the expense to produce a product for sale.
5. shows a company's net profit or net loss, the difference between revenues and expenses.

The Cash Flow Statement

The cash flow statement is represented by the following equation:

Cash Flow from Profits + Other Sources of Cash – Uses of Cash = Change in Cash

> The cash flow statement tells how much money was on hand at the beginning of the period, and how much was on hand at the end. It then describes how the company acquired and spent cash in a particular period. The uses of cash are recorded as negative figures, and sources

> of cash are recorded as positive figures. . . . The cash flow statement is useful because it indicates whether your company is turning accounts into cash—and that ability is ultimately what will keep your company solvent. Solvency is the ability to pay bills as they become due. (HBS, 2004, pp. 257–258)

Cash flow statements show how receipts and payments are directed into and out of, respectively, the company. Thus cash flow and profit are not necessarily the same. Cash receipts can come from a firm's sale of goods or services as well as interest revenue or dividend revenue. Cash payments can go to inventory purchases, payroll expenses, operating costs, or taxes. By contrast, investing activities often include purchase or sale of land, buildings, equipment, or other assets such as securities. Investments are usually not included as an operating expense as they are often indirectly related to the ongoing operation of the business. The final component of a cash flow statement involves financing activities. Financing activities that bring in cash include investments and issuance of company stock, bonds, or notes. Outlays of cash are payment of stockholder dividends and loan repayment.

The cash flow statement

1. reports the movement of cash into and out of a business during a specified period of time.
2. is broken down into operational activities, investing activities, and financing activities.
3. includes, in operational activities, all transactions related to the operating income of a business, both receipts and payments (HBS, 2004, pp. 257–258).

Starting Your Own Business

Many horticulture students have dreams of starting their own business within a few years of graduating from college. To start your own business, you will need to develop a comprehensive and well-thought-out business plan. A business plan includes an income statement and a balance sheet. It projects sales of the business as well as any necessary start-up costs. The business plan should also show funding sources. Funding may come from the owner's personal savings or investments or from outside investors and loans from banks or similar full-service lending institutions. If you borrow, determine the amount of capital you need and available interest rates.

New business ventures have an array of one-time start-up costs that the new owner must plan and budget for. How much capital you need depends on costs for utilities, salaries for your employees, equipment investments, and so on. The monthly rent or lease payments as well as heating, air conditioning, water, phone, and other utility charges must be planned for. Investigate probable salaries, health benefits, taxes, licensing, permits, insurance, and related operating expenses. Through which medium do you plan to advertise and how much do you anticipate spending on marketing-related needs? What office-related expenses, including computers, desks, chairs, tables, lamps, and other furnishings, do you foresee incurring? Do you know how much it will cost to purchase pickup trucks, dump trucks, skid-steers, rototillers, chippers, and stump grinders? Have you factored in the current costs of brooms, rakes, digging spades, hoes, wheelbarrows, backpack sprayers, gas-operated blowers, power edgers, and string-trimmers? What does a commercial-grade, fiberglass-handle shovel cost? Not answering these questions imperils your new business before it even starts.

CONCLUSION

Developing an understanding of the financial needs of a business is extremely important. Business owners who do not know their current financial situation cannot set long-term goals. Balance sheets, income statements, and the like are essential in making financial decisions. A sample balance sheet (Form A), income statement (Form B), and cash flow statement (Form C) are included.

NOTE:

Manager's Toolkit: The 13 Skills Managers Need to Succeed, part of the Harvard Business Essentials series, was instrumental in writing this chapter. I strongly recommend it to recent graduates entering the Green Industry as a well-written financial text.

FORM A

NATURALSCAPES DESIGN-BUILD SERVICES, LTD.
BALANCE SHEET
DECEMBER 31, 2006

ASSETS	
Cash	$85,000
Accounts Receivable	15,000
Inventory	50,000
Machinery	58,000
Vehicles	40,000
Land	95,000
TOTAL ASSETS	**$343,000**
LIABILITIES	
Accounts Payable	$60,000
Notes Payable	70,000
Mortgages Payable	80,000
TOTAL LIABILITIES	**$210,000**
OWNER'S EQUITY	
Capital	$133,000
TOTAL LIABILITIES & OWNER'S EQUITY	**$343,000**

FORM B

NATURALSCAPES DESIGN-BUILD SERVICES, LTD.
INCOME STATEMENT
PERIOD ENDING DECEMBER 31, 2006

REVENUE	
Income from Operations	$825,000
Interest Income (Bank)	400
TOTAL REVENUE	**$825,400**
EXPENSES	
Salaries	$250,000
Rent	25,000
Advertising	3,000
Cost of Goods Sold	500,000
Utilities	20,000
Truck Expenses	15,000
Miscellaneous Expenses	1,500
TOTAL EXPENSES	**$814,500**
TOTAL PROFIT	**$ 10,900**

FORM C

NATURALSCAPES DESIGN-BUILD SERVICES, LTD.
CASH FLOW STATEMENT
PERIOD ENDING DECEMBER 31, 2006

SOURCES OF FUNDS (CASH RECEIPTS)	
BEGINNING CASH	**$125,000**
OPERATING	
Sale of Goods & Services	600,000
Interest Revenue	5,000
Dividend Revenue	8,000
INVESTING	
Sale of Plant Assets	100,000
Sale of Retail Business	200,000
Sale of Other Held Business	50,000
Collection of Principal on Loan	2,000
FINANCING	
Issue of Stock	100,000
Issue of Bond	50,000
AVAILABLE CASH	**$1,240,000**
USE OF FUNDS (CASH PAYMENTS)	
OPERATING	
Inventory Purchase	(400,000)
Payroll	(100,000)
Operating Costs	(70,000)
Taxes	(10,000)
Interest Payment	(5,000)
INVESTING	
Purchase of Plant Assets	(100,000)
Purchase of Equity Securities	0
Loan Payments	(10,000)
FINANCE	
Dividends	(15,000)
TOTAL CASH OUT	**(710,000)**
NET CASH FLOW (Beginning Balance of Next Period)	**$530,000**

DISCUSSION QUESTIONS

1. What three concepts are critical to the long-term success of a business? Why are they important to the success of any Green Industry business?
2. According to the 2005 survey, the average annual revenue within the landscape industry was nearly $800,000 and the average net profit increased 11 percent over the previous year. What would you have expected it to be?

3. The biggest concerns landscape companies had for 2006 were (1) fuel prices, (2) worker compensation costs, (3) lowball competitors, (4) health insurance and overworked or overstressed employees, and (5) inflation and rising interest rates.

   Are these concerns still the most relevant in the landscape industry today? What concerns can you add? Have any lost their relevance?
4. What upcoming concerns loom on the horizon?
5. What reasons can you think of for contractors to spend so much more on subcontractors?

   Do you envision that such percentages will increase or decrease over time?
6. Why is it so important for companies to never neglect their customer service responsibilities?
7. Why should business owners in the Green Industry maintain an intimate understanding of their customer base?
8. Why is it important to proactively serve a dedicated core of customers?
9. Give some examples of ways to cater to the special needs of your customer base.
10. Define a rolling budget, an incremental budget, and a zero-based budget.
11. If you were starting a new landscape business, which budgetary process would you choose and why would it best suit the financial needs of your new company?
12. What constitutes the cash flow of a company?
13. Why are cash flow and profit not necessarily synonymous?

Ted Richardson's Experience in Starting His Own Full-time Business

Ted Richardson majored in ornamental horticulture at a large university in the Midwest and graduated with a 3.25 overall GPA. After graduation Ted joined Matsumoto's Landscape & Design as a crew leader. Matsumoto's was operated by Joe Matsumoto, whose grandfather had founded the firm over 40 years ago, building it into a sound and well-respected business. Joe had assumed the prime leadership 8 years ago, but he had other interests and did not have passion for the business. Ted had planned to remain with Matsumoto's for at least 4 years before starting his own landscape business, but during his 3rd year with the firm, he began to strongly consider starting his own landscape company sooner.

Joe Matsumoto possessed a number of favorable attributes, including honesty, integrity, a keen intellect, and a sincere love of plants. Joe had been very generous to Ted, especially regarding the young supervisor's overall compensation package. Ted enjoyed a more-than-competitive salary with the promise of year-round employment. Matsumoto's paid 100 percent of Ted's health-related benefits, including vision and dental coverage. Ted's 401k retirement plan was compensated at a generous 10 percent of his base salary, with a 50–50 shared match between Joe and his employer.

By September of his 3rd year with Matsumoto's, Ted had decided to start his own business and informed Joe Matsumoto of his decision. Ted already had a full-size, four-wheel-drive, heavy-duty pickup truck, which he had purchased used during his senior year in college. The truck was in excellent condition when he bought it and had been driven less than 15,000 miles by the previous owner. Ted's parents had made a generous down payment on the vehicle as his college graduation gift. Ted had earned a significant amount of money working as a landscaper throughout his junior and senior years at the university. His side jobs while in college and then his full-time salary at Matsumoto's had allowed Ted to pay off his truck loan. Ted had accumulated an ample supply of wheelbarrows, rakes, shovels, hoes, and other horticulturally related tools. Unfortunately, because all of Ted's personal finances had been spent on paying off his truck and purchasing landscape-related equipment, he had little more than $4,500 in his savings account. Ted planned to use this money to start his fledgling business.

A few days after Ted gave notice, Joe Matsumoto met with his young crew leader and told him that he believed Ted had significant potential as a self-employed business owner in the Green Industry. Joe said that he planned to soon pass the family business on to his younger brother and develop a B&B wholesale nursery specializing in high-quality trees and shrubs. Growing and marketing specialty nursery plants had always been his passion, and it was past time for him to more fully realize his personal and professional goals. He then strongly urged Ted to consider working with him in his nursery venture. Joe had a healthy amount of capital to invest in his new business. Joe told Ted that he would increase Ted's current salary if he would go in with him in a B&B wholesale nursery. What's more, Joe promised to share his financial expertise with Ted to better prepare him to start his own company. As they wrapped up their meeting, Ted sincerely thanked Joe and promised to seriously consider his offer.

After thinking it through, Ted decided to take Joe Matsumoto up on his offer of helping to develop a B&B wholesale nursery. Ted realized that gaining practical experience and saving additional capital would more fully support his long-term professional aspirations. Ted worked with Joe Matsumoto in his

B&B wholesale nursery for 4 years before leaving to start his own wholesale nursery business.

QUESTIONS FOR DISCUSSION

1. How much money should Ted have saved before starting his own landscape company?
2. Do you believe that Ted had accumulated enough practical, hands-on experience to consider starting his own full-time landscape business?
3. Did he have enough business experience to successfully operate his own landscape company?

   What advantages would Ted have gained if he had worked with a second established design-installation company before starting his own full-time landscape business?
4. Approximately how much would it have cost Ted each month to buy his own health care plan and save for retirement? How much did he save by having his employer pay for these?
5. Was Ted wise to pay off his pickup truck instead of saving the money?

   When paying off a bank loan on a top-quality, preowned, four-wheel-drive, heavy-duty pickup truck, what percentage of the monthly payment goes toward the loan interest versus the principal?

   What is the interest rate on savings accounts?
6. If you were a loan officer at a full-service banking institution, how highly would you rank Ted with regard to his ability to repay a small-business loan of $50,000?
7. As a loan officer, how would you evaluate Ted's ability to repay a small-business loan after he had worked with Joe Matsumoto at the specialty nursery for an additional 4 years?

CRITIQUE

Ted's collegiate accomplishments and 3.25 overall GPA made him a solid student academically. That he was a motivated and industrious young man is supported by his successfully completing numerous side jobs while in college. In addition, Ted had been diligent in paying off the hefty loan on his pickup. Despite these many strong points, he had weaknesses. His business experience and nursery knowledge were lacking, which could adversely affect his ability to operate his own landscape business. Ted had initially planned to start a landscape business but enjoyed working in the nursery industry so much that he made that the focus of his career. What does this say about Ted's familiarity with the Green Industry and his awareness of his own preferences?

We have several factors to consider when evaluating Ted's initial decision to leave Matsumoto's Landscape & Design and start his own landscape company. First, although Ted had paid off his truck loan, he had less than $5,000 in his savings account. Ted's plan to leave Matsumoto's in early autumn was perhaps not wise for additional reasons. Ted's new business would probably have been fiscally solvent through the end of October. Depending on weather, Ted might have earned sufficient money by completing landscape clean-up and maintenance jobs through the end of November. But winter business can be slow in the landscape trade. Although snow-plowing contracts can tide a business through to spring, such work is hard on vehicles, completely weather dependent, and sought by many with a plow and a four-wheel-drive truck. The hours are long and the pay is not as lucrative as one might expect, especially when vehicle repairs and similar costs are taken into honest consideration. Moreover, Ted would have been responsible for paying for his health and retirement benefits. Remaining fully employed and earning a healthy paycheck at Matsumoto's throughout the winter months looks better and better the more we look at it. Ted had failed to consider how little experience he had in evaluating budgets, balance sheets, income statements, and cash flow statements. All require a reasonable amount of business-related experience. Ted likewise did not have experience in hiring staff or in developing a trusted and reliable customer base. He did not have any experience in creating a business plan and obtaining a small-business loan from a bank or similar lending institution.

By helping Joe Matsumoto develop a wholesale B&B nursery, Ted would gain several business-related benefits. First, the managerial knowledge gained from such an experience would be invaluable to Ted in starting his own business. Second, it would be of immense value for Ted to witness firsthand how much investment capital is needed when starting a new business. The importance of saving more and borrowing less cannot be overestimated. Third, Ted would gain immeasurably from being mentored by Joe Matsumoto throughout the next 4 years. Joe was an honest and knowledgeable professional willing to share his considerable business knowledge with Ted. This would benefit Ted tremendously in helping him develop a business of his own.

Effective Meetings: Making the Most of Your Time

OBJECTIVES

After studying this chapter, you should have an understanding of:

- problem solving subordinate's needs
- delegating responsibilities
- reviewing progress of projects
- keeping abreast of work-related issues
- respecting time frames
- technical knowledge
- advisory personnel
- decision-making powers

Competent and capable Green Industry managers are very busy individuals and must make the best use of their time in meetings. Be early when meeting with your supervisor or other managerial staff. There are few if any exceptions to this rule. Arrive at least 5–10 minutes early for important meetings. This is a good habit to develop; it allows you to review notes and prepare for your meeting. There is little that appears more amateurish than a supervisor rushing in late to a meeting and keeping others waiting until he or she is ready to begin.

Understand beforehand the purpose of a scheduled meeting with a manager who is a direct supervisor. Some managers simply use this time to delegate additional jobs and responsibilities to staff, allowing little opportunity for guiding or mentoring a young supervisor. Astute managers use this time for learning about and problem solving subordinates' needs. They may also coach a junior supervisor through job-related problems. Most managers strive to combine these two extremes and use this time for (1) delegating new responsibilities, (2) reviewing progress of projects, and (3) keeping abreast of work-related issues that may need to be addressed. There could obviously be numerous other reasons for managers and staff to meet, including a review of budgetary issues, developing long-range departmental goals, or determining staffing requirements for the upcoming growing season. For the most part, however, meetings are normally used to solve pressing issues and monitor tasks of immediate importance. It is easy for the intended focus to change during the course of a meeting, but try to stay with the previously agreed-on agenda.

Be as efficient as possible in the time spent meeting with a direct supervisor. Address the most pressing problems before bringing up incidental issues. Your time together may be interrupted and such foresight assures that critical concerns are covered. Respecting time frames of meetings is a courtesy to all. If the meeting with your supervisor is scheduled to last 30 minutes, do not extend it to 45 minutes, even when it is informal and held on a pickup truck's hood. Always be extremely professional in any meetings you have with your supervisor. Be sure to use the time wisely and do not bring up issues that require only a simple e-mail. Finally, be prepared for the unexpected and do everything possible to foresee potential issues before they arise.

Here's an example of the consequences of a young supervisor not adroitly managing his meeting time with his manager. Max Hendry was the production manager at Rosetree Nursery's 375-acre wholesale B&B operation just outside Hickory, West Virginia. He was very well-respected in the trade and had been employed with Rosetree since graduating from college nearly 20 years ago. As Max and Denzel Hodges, his 2nd-year assistant, left the nursery's equipment barn at 7:30 that spring morning, Max instructed Denzel to meet him at the viburnums in section H at 11:00 a.m. By late morning, Max was well along in tagging a large quantity of 5–6 ft 'Winterthur' viburnums to be included in an order for a landscape and design firm. Max's objective was to finish tagging the remainder of the order, have a quick meeting with his assistant Denzel, and check on another of his nursery crews before lunch. It was the peak of the spring digging season and Max had many important responsibilities to attend to by the end of the day. When Denzel's company pickup truck pulled into section H, Max noted that his assistant was nearly 10 minutes late for their meeting. Denzel apologized for his tardiness, saying he had just left section D after checking on Randall's progress in filling a large order of 3.0–3.5 in. caliper Carolina silverbells and amelanchiers for Frost Brothers Landscape. Because Max was short on time, he told Denzel to tag the littleleaf lindens for another order and then check on Miguel's crew in section C. The skid-steer's hydraulic system had not been operating smoothly when Max had left Miguel a few hours earlier. As Max started to walk toward his pickup, he asked Denzel whether he was having any problems with the responsibilities he had been assigned earlier that morning. Despite having key information to give Max about crew delays, Denzel assured him that everything was fine and that he would check on Miguel as soon as he finished tagging the little-leaf lindens.

Max was obviously feeling the pressure of his duties on that spring-digging-season day, the busiest of times at a wholesale B&B nursery. Denzel was definitely remiss in being almost 10 minutes late for his 11:00 a.m. meeting with his supervisor. Pressures associated with filling spring digging orders make it especially important to be on time for even the most informal meetings. Time is a precious commodity at this time of year. Because the morning had been so hectic for Max, Denzel had decided not to mention that Randall's crew was behind in digging the Carolina silverbells and amelanchiers in section D. One of Randall's key crew members had arrived nearly 2 hours late for work that morning, and Denzel knew that Max was not aware of this. Denzel was wrong to decide to wait until lunch to inform Max that the Frost Brothers order would not be completed by the end of the day. It is imperative that young managers be totally honest with their supervisors about critical problem or issues that arise during the workday. Denzel also failed to inform Max that they were getting perilously low on 30 in. squares of digging burlap. He had planned to mention the need to reorder burlap but forgot in the rush of their meeting. During hectic times, when many deadlines loom, it is imperative to accurately convey important pieces of information so that nothing is left to chance. Denzel needs to work on being more reliable in communicating important details to his supervisor. When things are hectic, he should write down pressing issues as they arise to be sure he is prepared for his meetings. ❧

Getting the Most from Meetings

Inefficient meetings can be a major time-waster for any organization. To make them worthwhile, each meeting should be purposefully planned around a timely and important issue. Unfortunately, too many meetings occur because they are on the schedule and not because they are needed to deal with specific situations. They should be run efficiently. Time is a finite commodity, and once lost it can never be regained. Organizational time is an especially precious commodity if the business is to be profitable and successful. Each meeting's intent should be clearly defined and relevant to specific issues. Each attendee should be fully apprised of the meeting's objective and agenda. Consider using e-mail, conference calls, and other methods of communication as an alternative to gathering everyone in a room. Managers must constantly seek the most effective method for dealing with a specific problem. The more meetings scheduled, the less time supervisors have to manage their work-related responsibilities. Returns from meetings diminish the longer the meeting goes on. After 25–30 minutes, people's ability to focus on a problem is significantly reduced. Make sure that all in attendance support the final decisions and understand their responsibilities associated with follow-up meetings (Grossman & Parkinson, 2002, pp. 145–157; Figure 9-1).

When a meeting is scheduled with your supervisor,

1. arrive at least 5 minutes early.
2. be well prepared.
3. respect the meeting's scheduled start and end time.
4. discuss your most important issues first.
5. anticipate issues your supervisor may be interested in discussing.

One additional point should not be overlooked. Follow-up meetings should be as productive and well coordinated as the first. Numerous meetings scheduled to deal with the same or similar topics tend to deteriorate rapidly in their effectiveness. Problems must be solved and not simply dragged out over successive meetings with no forthcoming solution. Deal with an issue and move on.

Planning the Meeting

Significant planning is necessary to lead a successful meeting. Such planning demands forethought on how to attain meeting-related objectives. For example, be selective in choosing which staff should attend. Depending on the meeting's anticipated length of time or coverage of confidentiality issues, it may be beneficial to divide the meeting into two separate segments. This allows some staff to be excused from attending the second meeting.

FIGURE 9-1 A manager's weekly meeting with employees

To plan and orchestrate the most effective meeting possible,

1. make sure that the meeting is necessary in the first place.
2. take time to think through an agenda.
3. establish clearly defined goals.
4. do not stray from the agreed-on agenda.
5. clearly address legitimate concerns.
6. make sure that meetings begin and end on time.
7. strive to get consensus to solutions.
8. explore all significant options.
9. look for ways to successfully compromise and negotiate.

There are many types of meetings. Some meetings are necessarily quite formal, but others are best kept informal. Each type of meeting has its benefits, and you should select the format that best suits your needs. Keep meetings on schedule. Poor time management during meetings is perhaps the prime reason for their ineffectiveness. Prepare for any meeting you attend. Review all materials that are pertinent to your participation. Think through the reasons for inviting staff to a meeting and invite only those whose input is needed.

Young supervisors often invite only individuals with needed technical knowledge. However, inviting other advisory personnel can be advantageous. Individuals from another department can serve in an advisory capacity. Having an advisor with organizational decision-making powers who supports your goals will make their attainment more likely.

More practical considerations when planning a meeting include the following: Avoid holding a meeting immediately after lunch when the energy level of participants will be low. Similarly, schedule meetings to run no more than 30 minutes to prevent participants from becoming too lethargic. If it is July or August in a warm climate, be sure the meeting room is air conditioned. Consider holding a meeting off site to minimize interruptions from the daily shuffle of activities. Some meetings demand that refreshments and restroom facilities also be coordinated. If refreshments are served, attempt to address specific dietary needs. If the meeting will use audiovisual aids, be sure you know how to operate them. Supply sufficient copies of documents and reports for all attendees. And last, be sure the meeting reconvenes promptly after breaks or intermissions.

Your Professionalism Is Important

Planning an important meeting can be intimidating for young supervisors, especially if senior staff will be present. It will help if you remember to (1) be prepared, (2) be presentable, and most of all, (3) be confident. To be prepared, make sure information you present is up to date and accurate. If you are defending a request for new mowing equipment, be sure to get up-to-date price quotations. You don't want to try to convince the business manager of the need for a certain model of 72 in. cut, commercial-quality mower, only to have him or her tell you that the model was discontinued more than 18 months ago. Worse yet, what if the business manager finds that its price is $3,500 more than your cost analysis indicated? Your accuracy in financial projections and the research and documentation effort you put forth are part of your reputation.

Have a presentable, professional appearance in meetings. A simple, informal meeting may find you reviewing landscape plans with a crew member over the hood of a company pickup truck. A professional appearance in this situation may entail no more than business-casual attire: a three-button golf shirt with a company logo, khaki pants, and steel-toe work boots. On the other hand, a meeting between you and the chief executive officer of a private country club may call for business-formal attire: for men, a sports jacket, pressed slacks, and dress shoes; for women, dress pants or skirt with a jacket or a conservative dress, and low-heel, closed-toe shoes.

The physical presence you project can influence the opinions of those you are meeting with. This includes the tone of your voice and your body language. Young supervisors who understand the importance of projecting confidence will have more success in most meeting environments—especially when giving a presentation at a meeting, where you are selling yourself. You must believe in yourself and the topic you are presenting; remain believable.

Details of a well-orchestrated meeting include

1. considering travel times and transportation connections when out-of-town participants will be attending.
2. scheduling a time that is most convenient for the majority of participants.
3. sending a written confirmation of the time and place.
4. making sure that attendees are aware of any premeeting responsibilities.
5. putting the high-priority items first on the agenda.
6. choosing a convenient and comfortable site for the meeting.
7. keeping interruptions to a minimum.
8. taking clear and concise minutes of the meeting.

Managing the Meeting

How you choose to interact with those in attendance can have a strong effect on the outcome of most meetings. Good manners are extremely important for all to follow. It is never acceptable to be condescending or disrespectful. Listen attentively to others and give them due respect when they voice their opinions. As the leader of a meeting or as an attendee, display good eye contact, show interest in the discussion, and do not interrupt when others offer their comments.

When listening to the concerns of attendees,

1. give your full attention to the individual speaking.
2. look encouraging and initiate positive eye contact.
3. avoid negative body language when others are talking.
4. negotiate a win-win solution whenever possible.
5. keep an open mind and try not to manipulate the opinions of others.
6. quickly diffuse inappropriate outbursts of anger by other participants.
7. keep the meeting positive and aligned with your predetermined objectives.
8. stick to the facts and do not allow innuendos to infiltrate the discussion (Milo, 1989, pp. 82–83).

Not every meeting will flow smoothly. If problems arise, regardless of the situation, do not lower your professional standards. Some attendees may not have the professionalism to respect you and other attendees. There will be times when other participants do not share your opinions. Do not display outward signs of disagreement or distrust while they are speaking. Try to anticipate problems before they arise. Do not seat individuals together who may have a tendency to carry on private conversations at inappropriate times. Do not allow irrelevant discussions to disrupt the meeting. Never tolerate behavior aimed at disrupting the overall professionalism of a meeting. Some individuals tend to dominate meetings and inappropriately interject their opinions. Others may try to elaborate at great length on details irrelevant to the meeting's agenda. Whatever the behavior, the chairperson should manage the meeting as fairly and professionally as possible.

Concluding the Meeting

Many meetings are inefficient in their use of time and ramble on past the scheduled conclusion. Before concluding a meeting, be sure that everyone has had an opportunity to present relevant and constructive viewpoints. All points of business should be addressed and appropriate decisions made. The main points of the meeting should be summarized and the time and location of a follow-up meeting agreed upon. Outstanding items should be added to the upcoming agenda and related responsibilities assigned. Make arrangements for timely distribution of the meeting minutes to attendees.

CONCLUSION

Well-run meetings are an important component of operating a business in today's high-tech business world. Being able to orchestrate and participate in meetings at the highest professional level is often the hallmark of a mature and experienced supervisor. Without well-planned and productive meetings, it is difficult to achieve the organization's long-range goals.

DISCUSSION QUESTIONS

1. Why would it be advantageous to arrive 5–10 minutes early for meetings where your participation is expected?

   In addition to reviewing notes, what else might you do in preparation for a meeting when you arrive early?

2. What is the benefit to you of meetings with your supervisor that include an instructional and mentoring component?

3. Why adhere to a reasonably strict time frame in orchestrating and planning most business-related meetings?
4. Why be selective in choosing staff to attend a meeting?
5. What percentage of the participants invited to most business-related meetings would be more productive and valuable to the company if their time were spent elsewhere?
6. Why would you look for individuals to attend organizational meetings who (1) can provide technical expertise, (2) act in an advisory capacity, and (3) have decision-making power?

   In what ways does inviting an individual who can make final decisions add to the ultimate success of a meeting?
7. What might be the outcome of attending without preparation a meeting where financial information is to be discussed and presented?
8. With regard to being (1) prepared, (2) professional, and (3) confident, which of these three do you believe is the most important?
9. What do you gain by dressing professionally for important business-related meetings? Give some examples of meetings to which you would wear business-formal and business-informal clothing.
10. Why is it important to your long-term career to never be condescending or disrespectful to other participants at a company meeting? Fully elaborate your thoughts to this most important question.

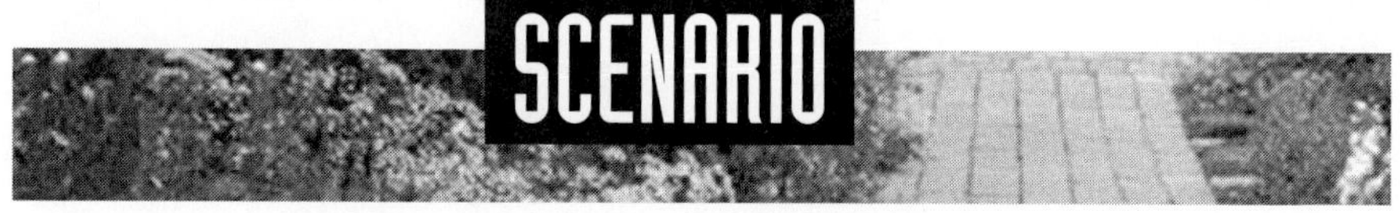

Lynn Kreiger's Budget Presentation at the North Hampton Botanical Garden

Lynn Kreiger had been director of the Horticulture Department at Long Island's North Hampton Botanical Garden for only a few months when she was given the responsibility of developing the department's budget for the upcoming year and presenting it to Mr. Thomas Raulston, vice president of finance for the garden. She had worked at the garden for 7 years. She worked there first as a crew member for 3 years and then as a crew leader for 4 years, before accepting the promotion to director of horticulture. She excelled in all her work and had deserved her promotions. She had graduated near the top of her university horticulture class with a cumulative 3.77 GPA.

Lynn looked at her upcoming meeting with Mr. Raulston as an opportunity to put her best foot forward as a relatively new supervisor.

Lynn understood that a budget defense was the protocol for all departments at the garden. Budgets assured that the institution was being fiscally responsible with its financial resources. Because the garden was still relatively new, each of the various departments would have significant needs to address. The North Hampton Botanical Garden was completing its 7th year, its grounds and mansion having been donated by the Armstrong family as a historical site to be transformed into a botanical garden.

Mr. Raulston called to suggest that he and Lynn meet at 9:30 a.m. on November 10 in the Quercus Room of the mansion as a first step toward solidifying the Horticulture Department's budget for the upcoming year. He told Lynn she could use whatever means she was most comfortable with to present and discuss her budget request. After sending an e-mail to Mr. Raulston confirming the details covered in the phone conversation, Lynn began planning how she could best use her time preparing for the meeting. Over the following weeks Lynn created a PowerPoint presentation listing the significant line items of the budget and a brief summary of why she made each request.

Part of Mr. Raulston's role was to determine which requests should be funded in the upcoming year's budget and which could be deferred to a later date. The two largest expenditures in the Horticulture Department's budget were a heavy-duty, four-wheel-drive pickup truck and a gas-powered cargo cart. The pickup truck was important as both a work truck and for plowing snow. The department's existing pickup truck had been used by the grounds staff when the estate was still the summer residence of the Armstrong family. The truck was now over a decade old and no longer suitable for day-to-day use. A gas-powered cart with a rear cargo bed was another important addition to the department.

The garden had no allegiance with a major manufacturer for cars and trucks, and so Lynn investigated three of the leading American manufacturers of full-size pickup trucks. She likewise obtained brochures illustrating specific technical features for three of the major manufacturers of gas-powered cargo carts. In her PowerPoint presentation, Lynn summarized in detail the technical qualities of each brand of truck and cart and rated the primary features of her top choices according to the needs of the department. Lynn obtained price quotations for her top two choices of heavy-duty, four-wheel-drive pickups and for her top two choices in gas-powered cargo carts. The information in her PowerPoint presentation was succinct, accurate, and easy to understand. Lynn highlighted each critical line item and summarized the remaining portions of her proposed budget: a third summer internship position for the department and a new computer system for Lynn's office. Lynn carefully quantified each significant budget request. A week before their November 10 meeting, Lynn provided Mr. Raulston with a detailed summary of the major points she would be making.

When the 10th arrived, Lynn made a well-documented and professional presentation. Mr. Raulston very much appreciated the thoroughness and attention to detail Lynn had obviously given to her presentation and complimented her on it. It was, however, his duty as the garden's vice president of finance to be fiscally responsible in managing the organization's monetary resources. Although Mr. Raulston did not question the need for any of Lynn's requests, he still had to balance her budget requests with those of the garden's other departments. He knew, for example, that the maintenance department had an equally dire need for a midsize backhoe and that the education department was requesting a large cargo van for their community-outreach programs. Dr. Lidia Radu, the executive director of the garden, was in the process of filling her secretarial-support position. Dr. Radu needed increased administrative support because of her fund-raising schedule and had requested that the position be changed to administrative assistant, with significantly expanded responsibilities. Mr. Raulston expected the position's annual salary to increase by $10,000–$15,000.

After discussing the Horticulture Department's upcoming budget for well over an hour, it was agreed that Lynn and the vice president of finance would reconvene in Mr. Raulston's office at 3:00 p.m. December 3. This would give Mr. Raulston approximately 3 weeks to review the budgetary needs of the garden and meet with each senior manager. He was to submit the final budget request to the garden's board of directors on February 7. In their meeting on the 3rd, Mr. Raulston again complimented Lynn on her professional and thorough presentation. Mr. Raulston told Lynn that he agreed with her need to upgrade her office computer system and she would have the money for that as soon as the new fiscal year began on April 1. Mr. Raulston had also authorized the hiring of the third internship position because of the obvious need for additional labor in caring for the garden's horticultural displays. He informed her that her request for a gas-powered cargo cart would have to be resubmitted during next year's budgetary process. As for the heavy-duty, four-wheel-drive pickup truck, Mr. Raulston understood that the existing truck could not be counted on to make it through another winter. He told her that one of the garden's newest board members owned a dealership for a brand Lynn hadn't included. The board member had promised to deeply discount a year-end model for the garden when the manufacturer's new models arrived in September. Deferring purchase of the Horticulture Department's pickup truck for another 9 months would allow Mr. Raulston to increase cash flow from membership sales and gate receipts. Although somewhat disappointed, Lynn was grateful for a new computer system, a third internship position and a new, although delayed, heavy-duty, four-wheel-drive pickup truck.

QUESTIONS FOR DISCUSSION

1. What was Lynn Kreiger's professional track record before she accepted the director of horticulture position at the North Hampton Botanical Garden?

2. Do you believe that Lynn had climbed quickly up the organizational ladder at the North Hampton Botanical Garden? Would you have been patient enough to spend 3 years as a crew member and 4 years as a crew leader before aspiring to the director of horticulture position at the garden?
3. Do you believe that Lynn successfully defended the budgetary needs of her department for the upcoming fiscal year?

   What could Lynn have done differently to secure the cargo cart in the current fiscal year?
4. Was Lynn correct in expending as much time and effort in preparing her department's budget for the upcoming fiscal year?

   Was she prudent in providing such an accurate and detailed presentation to Mr. Raulston?
5. Lynn made sure that her presentation was as succinct, understandable, and well presented as possible. How important are these characteristics in financially oriented information?
6. Lynn prepared for her budgetary defense for Mr. Raulston down to the most minute detail. Do you believe that her level of detail was appropriate?
7. Lynn provided Mr. Raulston with a detailed summary of the major points she would be supporting in their upcoming budgetary review a week before their scheduled meeting. Was this a wise move on her part?

CRITIQUE

The expectations on Lynn Kreiger in making the case for Horticulture's budget would have stretched the capabilities of almost any new supervisor, but Lynn did a masterful job of planning and presenting the defense of her department's budget. The entire weight of this most important responsibility had been placed on Lynn's shoulders, and it was her chance to shine. Her thoroughness and detail were in large part responsible for her securing a new office computer system, a third internship position, and a new heavy-duty, four-wheel-drive pickup truck. She didn't get everything she asked for, but managers are seldom awarded all of what they legitimately need when presenting their budget requests. Some might suggest she was too prepared for her meeting. Supervisors cannot spend an inordinate amount of time dealing with only one issue and ignoring their other responsibilities. The preparation and defense of a departmental budget is extremely important however. Those responsible for the fiscal health of an organization need to be extremely detail oriented. Mr. Raulston was most pleased with Lynn's efforts. Taking all these points into consideration, Lynn did a fine job in developing and presenting her departmental budget.

Mr. Raulston and the board of directors were undoubtedly influenced by the following in deciding what and what not to fund. Dr. Radu's administrative assistant would make the executive director more efficient in her fund-raising efforts. The Education Department's cargo van would support the department's many community education programs. Those programs would raise the garden's visibility in the local school systems and neighborhoods, which would eventually result in additional garden revenue. In the matter of Lynn's request for a new pickup truck, the brand she lobbied for was not sold at the new board member's dealership. Instead, she got a deeply discounted year-end model and had to wait several months to get it. No matter whether Mr. Raulston and the board were influenced by garden politics or other concerns, Lynn and the garden benefited from the board member's generosity.

The Maintenance Department got its midsize backhoe, but only because the need for it was known throughout the garden. Maintenance's superintendent was over a week late in submitting his department's budget for the upcoming year. He had made little effort to prepare anything resembling an appropriate budget request. On three separate occasions Mr. Raulston had requested that the maintenance superintendent provide him with a minimum of two price quotes for the backhoe. The superintendent eventually provided him with one price quotation. The superintendent's technical knowledge was more than adequate, but he was delinquent with a wide array of managerial responsibilities. Technical expertise is only one component of being a successful manager worthy of promotion. This superintendent's days remaining at the garden were numbered.

Planning: How to Be Most Effective in Your Position

OBJECTIVES

After studying this chapter, you should have an understanding of:

- setting of time frames
- prioritizing tasks
- flexible to change
- quantifiable steps
- evaluation process
- on time and within budget
- setting clear and focused goals
- career aspirations
- long-range planning

Planning is a necessary skill for managers in any organization. All facets of the Green Industry require planning, and planning is critical not just for the short term but for the long term as well. In fact, any attempt to plan strategically is useless if the organization has not established a solid foundation of long-term goals.

Components of Planning

Coordinating short- and long-term goals relies on the setting of time frames, estimated schedules for completing each job or task. Does the task need to be completed by noon today, by the end of next month, or by September of next year? Setting time frames in turn requires that you prioritize responsibilities.

Many managers use a simple A, B, C rating for prioritizing tasks. An A job has the highest priority and a C task has the least. This very basic method of planning fails to consider the time frame of the job in question and allows a task to remain unfinished for long periods of time. Take, for instance, a C priority that can be completed in 15 minutes today but will become an A priority 4–5 weeks from now that requires 12–15 hours to complete. It would behoove a supervisor to complete that task ahead of schedule.

Planning can never be divorced from delegation, scheduling, and communication. Along with the time frame for specific tasks, other critical aspects of the job must be considered. For example, who will cut the 4 acre section of turf along the entrance drive leading to the corporate headquarters and what mowers will be used. Moreover, the personnel and equipment needs of other supervisors in the company need to be considered.

Be flexible; whether dethatching the turf at a residence or applying a preventive fungicide to the 18th green at a golf course, planned tasks must be flexible to change and able to accommodate the equipment needs of another supervisor or a fast-approaching thunderstorm. In addition to being focused and well considered, a plan must have easily quantifiable steps and include an evaluation process. Was the turf mowed too short for the current drought conditions? Did the pin oak receive a technically proficient pruning within an acceptable time frame? Was the entire 18th green completely treated with fungicide despite an unexpected equipment failure? Was the foundation planting finished on time and within budget? Evaluation is critical to sound and effective planning.

When planning, a young supervisor

1. elicits constructive criticism from other trusted managers.
2. honestly evaluates the positive and negative aspects of a job-related task.
3. reprioritizes goals according to the changing needs of the organization.
4. coordinates short- and long-term goals.
5. develops a positive mind-set for reaching professional aspirations.
6. uses creative methods to attain work-related goals.
7. has above-average expectations for self and staff.
8. is firmly grounded in achieving quality-oriented results.

Putting Theory into Practice

A key component of successful planning is setting clear and focused goals. Usually, young supervisors are proficient in planning for the short term, but it takes considerably more experience and expertise to plan far into the future. Goal setting is critical to the satisfactory completion of important responsibilities.

As always, your overall goals must remain compatible with the organization's. Goals must be ethical; they should be specific. The goal of assuming a management position with Oceanview Landscape is too vague. Are you aiming to be a crew leader or the owner of the company? By the end of next year or before you retire? Managerial goals must be reasonable, not impossible. This can be more difficult than being ethical and specific. We all need to stretch ourselves in our career aspirations, but we should be realistic and mesh our true abilities with our professional objectives. A goal of owning and operating the largest landscape firm in the country might not be reasonable for a student who struggles to maintain a 2.13 GPA. Finally, professional or career goals are useless without impetus to complete them (Cairo, 1995, pp. 31–32).

Personal goals should mesh with professional aspirations. Priorities do change, and it would be unrealistic to expect that today's professional goals and aspirations would be identical with last year's, last month's, or perhaps even last week's. Establishing priorities keeps goals aligned. Major incompatibilities between personal and professional goals often result in a change of employment. For some, it may involve a simple move within the same general area of expertise. For others, it may mean an entire career change.

When planning,

1. strive to fit the many pieces of a puzzle into a coordinated whole.
2. remember that goals worth achieving are worth working diligently to attain.
3. heed respected managers and mentors.
4. keep the overall interests of the company in constant focus.
5. focus on the future and never dwell on the mistakes of the past.
6. learn from your errors.
7. accept responsibility for your actions.
8. understand that acceptance and support from others must be earned.

The closer you get to achieving a specific goal, the more focused and dedicated you need to be.

A Balancing Act

It is one thing to establish a to-do list; it is quite another to manage it properly. Some supervisors brag about the length of their to-do lists and relish that they have become so hopelessly overburdened. What they are forgetting is that it is not how busy a manager is but how productive. A to-do list is virtually useless if it makes you inefficient. If you're spending too much time determining what to do next, your to-do list is not working. Group related tasks so they can be managed most effectively. Make jobs boldface or italic or highlight in some other way those especially in need of timely completion. One of the most difficult aspects of being a manager is having to juggle many tasks. Managers need to be creative in the completion of their priorities. If your to-do list is overflowing with A-priority jobs, every one in dire need of completion, the first step is to reevaluate each task. Situations change and it is not at all uncommon for a job that was an A+ priority last week to warrant no more than a C priority today. The next step is to review your to-do list with an eye toward combining two or more related or complementary tasks. Consider delegating a few tasks, but be careful not to put them on the shoulders of someone who is already overburdened.

Planning requires using the most current information, using good time-management skills, and providing an acceptable level of quality. Most of all, don't procrastinate. If a job is an A priority and needs to be completed quickly, do it. Along the same lines, never make excuses for not successfully completing an important responsibility. This is a good lesson to learn from the beginning in one's effort to become a topflight and respected supervisor in the Green Industry.

How Not to Plan

Planning demands responsible use of an extremely precious commodity called time. Although you need to take care of business and remain responsible for your job, there will be occasions when you will be grasping for every uninterrupted second you can find. You cannot plan effectively if you are constantly battling every insignificant brush fire that comes your way. Your office telephone or e-mail can cause a great deal of time to be wasted. Some of the most productive time in a work-related environment is either early in the morning or late in the evening. The biological clock of some managers ticks strongest before the sun rises. For others, evening is when they are the most creative and problem solve to their best ability. Schedule planning time during periods you are least likely to be interrupted; these periods are unlikely to be available during normal work hours. The phone is ringing, meetings are scheduled, and employees need direction, correction, and instruction throughout the workday.

Another way to maximize your planning time is to do your most mentally taxing and intense work when your mind is in its most productive state. If you are not a morning person, do not schedule your crew's work before the sun rises in the morning. Less-critical planning can be done when your thought processes are not at their best. Find the planning method, often an electronic planner or computer, that works best for you. Much depends on the complexity of tasks and the resources that best fit your style of management. Long-term planning is often measured in years, and planning so far into the future is difficult but essential to do. Long-term planning is the backbone of any well-managed organization, especially within the often seasonal horticultural profession.

The flow of a horticultural business changes from month to month throughout the year. A garden center should have ordered all of its vegetable seeds, bare-root trees, shovels, hoes, rakes, pruning tools, and fertilizers well in advance of the spring planting season. A florist who has not placed an order for red, long-stemmed roses well before Valentine's Day will not reap the financial rewards of such an important holiday. A wholesale nursery needs to have wire baskets, burlap, and twine well before its digging season gets under way, and a golf course superintendent would need to stock fertilizers, fungicides, and herbicides far in advance of spring. Thinking even further ahead, the director of horticulture at a large botanical garden would be well-advised to plan for a new greenhouse or large equipment-storage facility a year or more in advance of anticipated need.

To plan effectively,

1. be confident and believe in yourself as a professional.
2. focus on the task at hand.
3. have a sincere appreciation for those who work diligently to support you.
4. instigate positive change through dedication and motivation.
5. show respect for your professional beliefs and commitments.
6. have a personal and professional desire to succeed.
7. maintain a clear commitment to quality.
8. respect the financial management of the organization.

CONCLUSION

Long-term planning is an educated attempt by managers to make business-related decisions based on their forecast of future events. For example, a recession might reduce a family's disposable income and force them to defer needed landscape improvements. On the other hand, job gains, tax incentives, and lowered interest rates would encourage families to install a new flagstone patio, a dry-laid fieldstone wall, or foundation planting.

None can accurately forecast the financial picture of their business a year or more into the future, but we must do our best. Many of the issues associated with sound financial planning relate closely to starting a business. If, for example, you have worked for 4–5 years in a managerial position with an established design-build landscape firm and start thinking about opening your own business, what questions would you need to ask? How much capital do you have saved and how much money would you need to borrow and at what interest rate? Where would you locate the business and what type of building or structure would you need? How long would it take for the company to become profitable? What would be the focus of the company? What level of experience and education would qualified staff need? What would be the responsibilities for each position and what salary and benefit packages would you offer? How many full-time staff would you have, and would you offer internships? How many seasonal staff would you hire and in which months? What are your direct and indirect business competitors? Should you have a business partner? Can you establish a dedicated base of clientele that goes beyond close friends and relatives and continue building that base? What advertising mediums would you use? What financial planner, accountant, and legal advisor would you employ? These are only a few of the questions you need to consider and answer.

DISCUSSION QUESTIONS

1. Why schedule smaller and lower-priority tasks for completion before tackling larger and higher-priority jobs?
2. Why assign a time frame for each task?
3. Why is planning so dependent on delegation, scheduling, and communication?

   Detail the importance of each to planning.
4. Why should managers be flexible in planning, and why should plans be quantifiable?
5. Is it possible to implement plans without being adept at both written and oral communication?
6. Why should personal goals mesh closely with career aspirations?

   Is this meshing of goals more important or less important in earlier stages of a career? Support your answer.
7. What does it mean that planning is a balancing act?

   Is it more true for experienced supervisors or young supervisors?
8. How do time-management skills aid planning?
9. What advantages do managers who know their biological clock have when planning?
10. Why is it important for young supervisors to have a sound knowledge of the month-to-month flow of a business, and is such an understanding necessary for the purposes of planning?

    Explain how the month-to-month flow in the different types of horticulture industries can be dramatically different from each other.
11. Describe how the business cycle for a garden center in South Carolina would be different from a garden center in Massachusetts or New Jersey.
12. How will your crew's ability to work together as a team make planning more efficient for you as a young supervisor?

Martha Butler and Her Work Experience at Dresher's Garden Center

Throughout college, Martha Butler's goal was to eventually operate her own garden center. During most of her summer vacations in high school and during the summer following her freshmen year in college Martha helped out at Dresher's Garden Center just outside Springfield, Massachusetts. Thomas and Anne Dresher, Martha's uncle and aunt, had operated their successful garden center business for 27 years. Martha had been given progressively more responsibility at the garden center and had by now become quite familiar with its overall operation.

Martha more than fulfilled the requirements of her college horticulture internship by working at Hargrove's Garden Center & Nursery during the summers following her sophomore and junior years at college. Hargrove's, located on the Hudson River approximately 60 miles north of New York City, was respected throughout the region. Hargrove's catered to middle- and upper-middle-class clientele and had developed a faithful following of dedicated customers. Their prices were decidedly above other garden centers in the region, and Hargrove's prided itself on providing high-quality merchandise and exceptional customer service. Martha's internships at Hargrove's provided her with a practical understanding of operating a garden center business. Martha greatly expanded her overall knowledge of woody and herbaceous plants in her first summer there. During the second summer, Martha was employed in the business office and worked very closely with the firm's senior business manager. She learned Hargrove's invoicing, ordering, and accounting procedures firsthand.

Martha's aunt and uncle, in their early 60s, had discussed with her the possibility of selling the business to her when they retired. After graduation she began working there full time, and her aunt and uncle encouraged her to work in all aspects of their business. Martha soon established a relationship with the garden center's long-time vendors and began to place seasonal plant and hardgood orders. She coordinated the purchase of the garden center's Christmas trees and related holiday items and helped oversee a large poinsettia crop in the greenhouse. Martha coordinated the development of a sizable number of high-quality decorative fruit baskets, a new item for Dresher's. To promote their sale, Martha and Mrs. Dresher converted a former storage room adjacent the main section of the garden center to display the fruit baskets and related holiday items. By sealing off the room from the main garden center and cooling it they maintained the quality of the fruit at a consistently high level. The fruit basket display was a significant success and received positive response from the many holiday customers who visited the garden center. Soon after the New Year, Martha met with Mrs. Dresher to coordinate the many vegetable and hardgood orders that would be needed for the upcoming spring season. After reviewing tool orders placed over the previous 3 years, Martha met with each hardgood vendor and ordered all the rakes, shovels, hoses, and related supplies needed for the spring season. She ordered lawn, garden, and household fertilizers and fungicides, insecticides, and herbicides. Martha also significantly increased inventory of organic plant health products and coordinated the display of the merchandise as it began to arrive for spring sale.

The Dreshers had always sold a wide array of annual flowers and vegetables, but they had not investigated herbs and related plants for their more discriminating clientele. Martha saw this as a missed opportunity and got Mrs. Dresher's approval to expand this aspect of the operation. In addition, Martha realized that customers with large home vegetable gardens would want more than the smaller, commercially available seed packets the Dresher's and other garden centers had sold in previous years. Martha had various varieties of sweet corn, bush beans, pole beans, peas, and other seeds bagged into larger quarter-pound, half-pound, and 1-pound quantities.

Martha gave much thought to ways the business could cut costs and still maintain the highest level of quality and service. She suggested approaching other privately owned garden centers in the area to collectively order a wide variety of hardgood products. At the very least, this would allow the various businesses to save on product and shipping costs. Bales of peat moss, bags of topsoil, insecticides, fungicides, pre- and post-emergent herbicides, and blends of fertilizers could be purchased in bulk and directly shipped to a central location for redistribution to each garden center. The Dreshers thought the idea had merit, seeing that each small, privately owned garden center could then better compete with the often lower prices of larger chain-store operations. After much brainstorming, Martha and the Dreshers set up a meeting with several competitors to discuss the program. Later that winter, Martha placed an order for Dresher's and nine other neighboring garden centers. All of the participating garden centers benefited from the partnership and continued to meet quarterly and strategize how they could better keep their collective operating costs to a minimum.

Martha worked extremely hard over the next few years to assist the Dreshers in growing and expanding their business. She upgraded the overall quality of the operation and increased its profit margin. Martha's biggest success was developing a financially sound operation despite mounting pressures from chain-store competition. As she continued to grow in the business, Martha never lost sight of the importance of providing the highest level of customer service to her loyal customers. During the remaining years of helping her aunt and uncle, Martha saved a considerable amount of capital as a down payment toward purchasing the business. As her aunt and uncle prepared to retire,

Martha applied for and secured a low-interest loan from a local bank to purchase the business.

QUESTIONS FOR DISCUSSION

1. What sort of experience did Martha gain toward owning her own garden center by serving the demanding clientele of Hargrove's Garden Center & Nursery?
2. How did Hargrove's excellent professional reputation benefit Martha's growth and development as a future small-business owner?
3. What do you think Martha could have done to additionally prepare herself for assuming ownership of her aunt and uncle's garden center?
4. In addition to the sale of holiday fruit baskets, what other ideas or improvements could Martha have instituted in the operation of the Dreshers' business?
5. After uniting the area's privately owned garden centers in collective purchases of hardgoods, what else could Martha suggest to do along those lines?
6. In today's dollars, how much capital would be needed for a down payment on an existing garden center? Consider the region of the country, location of the garden center, size of the business, and so on.

   If a recent graduate opted to start his or her own garden center, would the investment be larger or smaller than for purchasing an existing operation?
7. What percentage of the property's worth would a loan officer at a local bank require as down payment on a midsize garden center?
8. What is the average interest rate for a small business loan from a bank or similar type of lending institution in your region of the country?

   What is the average term of a small-business loan, and how is the interest compounded?

   Check at least three lending sources before answering this question.
9. Garden centers would need to stock turf fertilizers, post-emergent herbicides, potting soil, and sphagnum peat moss. Develop a further list of the major supplies that a garden center would require to operate.

   Develop a comprehensive list of equipment, such as pickup trucks and front-loading tractors, that a garden center would need to operate.
10. Assign realistic prices for new and preowned equipment in item 9's list. How do the prices between new and quality preowned equipment differ?

    What are the pros and cons of purchasing new equipment versus preowned?
11. How long do you think it would take before a mid-sized garden center could be expected to make a profit?

12. When a garden center, nursery, or landscape business fails and its equipment is sold at auction, what percentage of the market value will the preowned equipment likely sell for?

CRITIQUE

From her internship experiences at Hargrove's to the purchase of Dresher's Garden Center, Martha had a plan and executed it with vision and foresight. Let's investigate the finer points of Martha's plan and determine how she best coordinated her decision-making abilities with her long-term career goals.

Martha wisely obtained as much practical, hands-on experience as possible. She not only worked at Dresher's Garden Center in college but also completed two separate internships at Hargrove's. Although Martha could have worked exclusively at Dresher's throughout college, she stretched herself and worked for a very competitive garden center 60 miles north of New York City. Hargrove's was a fast-paced and demanding environment and provided two extremely positive educational experiences for Martha.

Martha was also wise to choose for her internships a garden center that allowed her to experience so many facets of its business operation. The opportunity of being trained by the firm's senior business manager was a priceless experience for a college student. She would not have had that learning experience at most garden centers. Martha's experience at Hargrove's gave her a firm understanding of a complex garden center operation, what was required to operate it, and how to cater to an extremely discriminating and demanding customer base.

Rather than interning at a different garden center or Dresher's, several factors led her to work again at Hargrove's during her second and final summer before graduation. Martha had had a very positive learning experience there her first summer. Martha discussed her options with Hargrove's following her first summer there and was assured she could diversify her learning experience if she came for a second internship. Hargrove's believed in Martha's abilities and that she had proved herself a dedicated and committed professional during her first summer with the firm. If she had interned at another garden center, not Dresher's or Hargrove's, it almost certainly would not have been as worthwhile as her second summer at Hargrove's. Martha could easily have gone back and worked at Dresher's during her final summer in college, but that would not have challenged her. The Dreshers were looking out for Martha's best interests when they encouraged her to intern with other garden centers. In the long run, the Dreshers benefited from Martha's internships with Hargrove's by gaining a much more seasoned and well-rounded employee.

Her decision to intern at a garden center in a northeastern state showed good judgment in choosing a company in the same general climatic region as was Dresher's Garden Center in Massachusetts. Pursuing an internship experience in Miami, San Diego, or Dallas–Fort Worth would not have given Martha the experience she needed to run a garden center in the northeast.

As a permanent full-time employee at Dresher's, Martha made significant improvements to the operation. A new perspective refreshes any small, family-owned business. After 27 years in the business, the Dreshers were probably not as innovative as they could have been. Martha's idea for the area's privately owned garden centers to purchase hardwood products collectively made them all more competitive with larger chain-store operations and lowered their operating costs. It was well received by the other privately owned garden centers and helped them support each other.

Martha showed maturity and prudence in saving a significant sum of money. Due to her fiscal responsibility, she had to borrow considerably less money and thus carried far less debt. Martha astutely planned and orchestrated her long-term goals and objectives.

Dressing for Success: Looking Your Part

OBJECTIVES

After studying this chapter, you should have an understanding of:

- the importance of first impressions
- coordinating professional dress with career goals
- coordinating professional dress with corporate image
- respecting corporate dress codes
- adapting professional dress to day-to-day circumstances
- professional dress vs. fiscal health of the organization
- fads and trends vs. professional image
- peer pressure vs. professional image

Choose your professional dress and outward appearance with three primary factors in mind. First, follow the organization's established dress code. Second, do your best to meet client expectations on appearance. Third, your dress and overall appearance should support your professional objectives.

Although your clothing can influence others' impression of you, your image goes well beyond attire. *Image* comprises the overall professionalism a manager displays daily, whether wearing a uniform shirt and khaki pants to check on the crew's morning progress or attending an afternoon meeting with the CEO of an exclusive country club. ❧

Representing the Organization

Your outward appearance should complement the overall goals and objectives of the organization. Dress-code standards are often given in the company employee handbook, and there is little reason for staff not to be familiar with an organization's dress code. Guidelines for appropriate dress or appearance are directed not only to supervisors and managers but to support staff as well. Inappropriate dress usually results from how a particular staff member chooses to interpret company directives. For example, if it is organizational policy for hourly personnel to wear khaki pants, a T-shirt displaying the company's logo, and steel-toe work boots while on the job, wearing dirty pants or a torn shirt is inappropriate. The ultimate purpose of a dress code is to encourage staff at all levels to convey a professional image.

A company dress code cannot foresee all situations. If the country club where you are the assistant grounds superintendent is hosting a nationally televised Ladies Professional Golf Association tournament, senior management may dictate special attire for staff employed at the event. The requirements for professional appearance will undoubtedly change as staff members grow in their level of responsibility. What may have been appropriate dress for a crew leader would not be at all suitable for a division manager or director of landscape services. As you mature to positions of higher authority, it becomes even more important to take overall appearance into greater consideration.

It is also imperative to understand that the responsibilities of professional appearance may vary significantly by situation. This can be especially true of those in sales-related positions. Although many horticultural companies have specific guidelines for their sales staff to follow, it is still important to appreciate the wants and expectations of the customer base. Selling a tree spade attachment to a rural B&B wholesale nursery would most likely entail its demonstration, and a comfortable pair of work pants and steel-toe leather boots are more suitable than a sports coat or blazer. By contrast, a sports coat may be the correct attire for a salesperson meeting with the owner of a high-end garden center to discuss retail possibilities for highly artistic, handcrafted lawn sculpture.

When meeting with an important client, make sure that your

1. hairstyle is appropriate and professional.
2. clothing fits and is sized appropriately.
3. clothing's colors are subdued and well coordinated.
4. clothing is in style and tastefully conservative.

Three main criteria help you decide what is appropriate. The first is simply to be guided by other successful managers whose actions and opinions you trust. The second is to understand the situation and use common sense. As an example, would you expect the manager of a surf shop in southern California to be dressed in a tie and sports jacket? Of course not. Finally, try not to allow fashion trends or peer pressure to influence your dress-code-related decisions. Some companies are fairly conservative regarding the professional image they want their staff to portray. Take note of the corporate mind-set, and follow established guidelines.

First Impressions

Attracting and keeping customers is the financial lifeblood of any organization. Your professional dress and appearance influence the purchasing decisions of clients, and first impressions are critical. Those who meet you for the first time make personal and professional judgments in less time than it takes to order lunch at a fast-food restaurant. In a few short minutes they evaluate

Those meeting you for the first time will within a matter of seconds judge your

1. overall intellect and level of education.
2. professional competence.
3. honesty and overall trustworthiness.
4. ethics and morals.
5. personality and interests.
6. motivations and intent.

your physical features and your intellect. By the time a few more minutes have passed, they decide whether you are honest and can be trusted. Before the meeting is over they will have made inferences as to your professional capabilities and overall level of integrity. Such judgments are not necessarily accurate nor necessarily fair. Unfortunately, first impressions can be all some people have to base their opinions on. This can be especially true in sales-related situations, where a client can be literally won or lost in a matter of minutes (Willingham, 1992, p. 9).

Changes in the Green Industry over the last few decades have affected the career choices of today's young professionals. Family-owned and family-operated businesses have for years formed the backbone of the horticulture profession. Although they will remain an important component of the industry well into the foreseeable future, recent graduates can entertain a much broader range of employment opportunities than ever before. Arboretums, botanical and zoological gardens, theme parks, and estate horticulture often bring with them national if not international reputations. Because so many are managed as entertainment venues, they are often much more structured in their dress code and the appearance they demand of their staff. High expectations are evident also in other sectors of the industry, and it is imperative that young supervisors appreciate how their professional appearance can affect their future.

John Cleans Up

John Witherspoon was an arboriculture crew leader with Folcroft Tree Experts, based just outside Boston, Massachusetts, and with operations also in Ohio, New York, Pennsylvania, and New Jersey. John had started with the company directly out of college and had now been employed with Folcroft's for 7 years. He had started as a climber and then been promoted to his current position of crew leader after little more than 3 years. He had been a crew leader for over 4 years now and had been passed over for promotion three times in the last 2 years. During all his time with Folcroft's, John often came to work with his shirt and pants wrinkled and soiled. John's work boots were in even worse condition, the leather having been worn off one toe and exposing the protective steel plate beneath. John had applied 4 months ago for a sales position with a firm just outside Pittsburgh where he would have been responsible for corporate and residential accounts. The position was filled with an in-house candidate who had fewer years of experience than John,

As John drove home from work in his mud-splattered pickup truck one day, he stopped by the regional office of a large and highly respected tree service company. He inquired whether the company was hiring full-time employees and was told no by the secretary he spoke to. Despite his mentioning that he had an arboriculture degree from a respected 4-year university and had 7 years of work-related experience, she seemed unimpressed. Could it have had something to do with his appearance? Over the next few weeks, John couldn't help but notice the company's employees working at job sites throughout the area. Each employee always looked well-groomed and presentable. John noticed too that their trucks were always washed and their equipment was well maintained and cared for.

John decided to get a haircut and took his shirts and pants to a dry cleaner to have them professionally laundered. Each piece of clothing came back clean and wrinkle free. John also purchased a new pair of work boots and stopped by the car wash with his pickup truck. In the end, John himself was impressed with his appearance. A few weeks after his first visit, John returned to the regional office. He entered as the general manager was finishing a conversation with one of his supervisors near the front desk. As John inquired about current employment opportunities, the general manager overheard and struck up a conversation with John. They conversed for nearly 20 minutes. A week later, John was called to come in for an interview and was awarded the position of sales associate with the company.

What can we learn from this? John was fortunate to be in the right place at the right time to have a conversation with the general manager of a regional office about job opportunities. However, if John had not cleaned up his appearance, it would not have made any difference whether the general manager was present. The better companies seek only first-rate employees. Although there are admittedly many qualities that make up a truly top employee, professional dress and grooming are often among the most important. This is especially true for service-related professions that rely on repeat business from customers. Landscape contractors, arboriculture companies, garden centers, and florists all represent profit-driven companies whose financial well-being depends on a satisfied base of clients. This is particularly true at high-end country clubs where members annually pay $20,000 or more. Nonprofit organizations such as botanical gardens or arboretums also depend on well-groomed and neatly attired employees. Nationally known theme parks can also be quite stringent in their dress code and grooming requirements. Given that many Green Industry professions are mud-on-the-boots occupations and that young managers are working supervisors, keeping a clean change of clothes behind the seat of the company pickup truck is one simple way to help assure a positive first impression to a client or customer.

When meeting another for the first time,

1. offer a firm yet controlled handshake.
2. make sincere and positive eye contact throughout the encounter.
3. avoid making any distracting gestures.
4. do not monopolize the conversation.
5. avoid appearing nervous, especially if the individual is well-known.
6. make sure that you end the conversation on a positive note.

CONCLUSION

Few positions in the horticultural profession demand as high a level of professionalism and overall presentation as those at a high-end resorts, theme parks, or nationally televised golf events. In each of these venues, quality remains the overriding focus. Whatever aspect of the Green Industry you choose to pursue, always keep in mind that a young supervisor's professional dress and appearance are often key to his or her long-term success.

DISCUSSION QUESTIONS

1. In addition to predicating professional dress and outward appearance on established organizational guidelines, the expectations of customers, and your career goals, can you suggest anything further that should be considered?
2. Professional appearance goes well beyond the care given to choosing his or her dress or attire. What additional factors are important?
3. Why should young supervisors care about their professional attire and overall appearance, and what does that have to do with the organization's interests?
4. Give some examples of situations for which an organization's employee handbook will not provide guidance for proper dress or appearance.

   How much does good judgment and common sense play into this decision-making process?
5. If you owned a landscape business, how would you expect your staff to project a positive and professional company image?

   How important would that be to the company philosophy?
6. What benefit might there be for subordinate staff's appearance from holding management to higher standards regarding their professional dress and appearance?
7. If you represented a horticultural hardgoods supplier and traveled to garden centers to sell herbicides, fungicides, fertilizers, terra-cotta pots, commercial quality tools, and similar high-quality items, how should you dress?
8. If you represented a horticultural book publisher and were responsible for setting up large, visually attractive displays at heavily attended metropolitan flower shows, how should you dress?
9. If you represented a nursery supplier and were responsible for on-site visits to rural, wholesale B&B operations throughout the mid-Atlantic states, how should you dress?
10. If you were the customer-service representative for a landscape firm that specialized in serving upscale and discriminating clients, how should you dress?
11. Give some examples of how dress and professional appearance can have a significant impact on your ability to provide first-rate customer service.
12. How long does it take you to make judgments about intellect, character, and overall professionalism when you meet someone for the first time?
13. If you were the head turf superintendent at a high-end country club, would overseeing your staff's commitment to maintaining a professional appearance be an important component of your supervisory responsibilities?

James Lyden's Managerial Experience at the Marshall Residence

James Lyden was employed as a crew leader with Peterson's Tree Service in North Newbury, Connecticut. Peterson's was a long-established arboriculture firm started by John Peterson's grandfather in the mid-1950s. Peterson's had earned a reputation for high-quality work and exceptional customer service. The firm had not advertised in many years and received all of its business as a result of referrals from long-satisfied customers.

James had graduated with a degree in arboriculture from a well-known university in the northeast. He had worked as a college intern with Peterson's and impressed the firm with his climbing skills and overall technical ability. James started full-time with Peterson's directly after graduating from college and was promoted to crew leader at the beginning of his 4th year with the firm. Now well into his 5th year as a full-time employee, James had recently been given a second crew to supervise.

Bo Johnson, hired by Peterson's approximately 5 weeks ago, had worked as a climber for numerous arboriculture firms throughout his 15-year career. Although he had displayed excellent technical abilities as a climber, the firm had begun to question his overall character and customer-service skills. Over the past few weeks, Bo had worked under the supervision of Chun Hei Yee, a trusted and highly responsible manager who had been with Peterson's for over 20 years. Chun Hei Yee had been involved in a minor motorcycle accident the previous weekend and had not yet returned to work. As a result, Bo was working this morning at the Marshall residence under James's direction.

James could sense when giving out the morning's work assignments that Bo did not care to be put under the young supervisor's direction. Bo had been asked to go with two other experienced employees to thin the crowns of two midsize pin oaks at the Marshall residence. The Marshalls, long-time customers of Peterson's, had planted a wide array of semi-shade-loving herbaceous perennials under three midsize oaks. One of the three oaks had been thinned by Peterson's the previous year, and Mrs. Marshall wanted to expand the current understory planting to include a broader palette of perennials. As a result, she had requested that Peterson's return to her property and lightly thin the two remaining trees.

As James pulled his company pickup truck into the Marshall's driveway just after 1:45 p.m., he could not hear a chain saw running at the rear of the home. As he walked around the garage, he noticed two of his employees finishing the lawn cleanup in the vicinity of the oaks. He was impressed with the quality of the work that had been accomplished. James noticed that Bo and one of the company trucks were absent. When asked about Bo's location, one of the employees shrugged his shoulders, saying that Bo had left for the hardware store about 11:00 a.m. and not returned.

About that time, Mrs. Marshall came out to compliment James on the work of his employees. He was dressed very professionally and was wearing his supervisory uniform of tan work pants along with a forest green, three-button short-sleeve shirt complete with an embossed company logo. James had supervised much of the work at Mrs. Marshall's in the past, and she had always been impressed with the professionalism and respect he showed her. Hearing James inquire about Bo's location, Mrs. Mitchell mentioned that she had noticed a Peterson's Tree Service truck parked in front of Casey's Bar & Grill about a mile down the road. James did not frequent Casey's, but he knew it did not serve lunch. He thanked Mrs. Mitchell for the information, apologizing on the firm's behalf for any inconvenience and thanked her for her continued support of Peterson's Tree Service.

After checking on the progress of his two remaining employees, James called Mr. Peterson on his company cell phone to explain the situation. Mr. Peterson said he would handle Bo. He asked James to tell Mrs. Mitchell that she would not be charged for Bo's time at her residence, an additional 10 percent would be deducted from the invoice in appreciation for her understanding of the situation, and she would not be charged for the extra time that it took his remaining two employees to complete the job.

QUESTIONS FOR DISCUSSION

1. Do you believe that James had anything to do with Bo's inability to act and perform professionally, or do you think that Peterson's Tree Service simply made a misguided decision in hiring Bo for full-time employment?
2. After Bo's departure, only one experienced climber was left to thin the pin oaks at the Mitchell residence. What safety concerns arose at the Mitchell residence as a result of Bo's leaving the work site?
3. Even if Bo had not gone to Casey's Bar & Grill, was it still a serious breach of professional responsibility for Bo not to inform his fellow employees when he planned to return?
4. After parking the company pickup truck in front of Casey's Bar & Grill and not returning to the job site after lunch, do you think that Bo had any intention of returning as an employee with Peterson's Tree Service?
5. Do you agree that Mr. Peterson made a wise decision not to have James enter Casey's Bar & Grill in an effort to locate Bo Johnson?

6. Do you believe that James was positive and proactive in helping to manage the overall situation at the Mitchell residence on behalf of Peterson's Tree Service?
7. What do you think of Mr. Peterson's handling of the billing and invoicing issues?

   Do you agree with Mr. Peterson's decision that at times small-business owners need to be generous with their long-term clients even though a particular problem may have been well beyond the control of the company?
8. Why does a company vehicle have no place being parked in front of a bar at any time and for any reason?
9. Is it a difficult task in any facet of the horticulture profession to hire quality individuals who will represent a company with the proper degree of integrity and professionalism?

CRITIQUE

Before reviewing the events at the Mitchell residence, let us consider a few alterations to this scenario. Suppose Peterson's had been not a third-generation arboriculture company but a new company in business for only a couple of years. Suppose Mrs. Mitchell had not been a long-standing customer but was a less than cordial and understanding client. The damage control required for this latter situation could have been much worse. This would be especially true if the newly formed business were operating in a small community environment, where a firm's reputation can be much more easily damaged.

Mrs. Mitchell had understood the implications when she saw the Peterson's Tree Service truck parked at Casey's Bar & Grill. James could not cover up the situation, and fortunately, he did not try. James was totally honest with Mrs. Mitchell, and she appreciated his candor and professionalism. James obviously represented himself very well on behalf of Peterson's Tree Service. He was well-groomed and polite, courteous, and totally attentive in his interactions with Mrs. Mitchell. James did an outstanding job of evaluating the overall situation and coordinating with her.

With regard to retrieving Bo from the bar, it clearly would not have been a good decision for James to enter the bar and confront Bo. At that point, Bo was obviously going to be terminated and it was really just a matter of getting the pickup truck back to the company's yard. Furthermore, it had been obvious that Bo did not appreciate interacting with a much younger supervisor and, by that time of the afternoon, he had been drinking for hours. There was no need, in Mr. Peterson's mind, to make a bad situation worse. His company's reputation would only suffer as a result.

Broader issues were at stake with a company truck parked in front of Casey's Bar & Grill. It was a small-town environment, and this could only be negative advertising, especially when the vehicle was there in the middle of a workday. Company vehicles should never be seen by potential or existing clientele in the parking lot of a bar, including company vehicles that employees take home after hours.

Mr. Peterson was much more interested in providing satisfactory service and retaining a long-time client than making a significant profit on the thinning of the Mitchell's two pin oaks. He did not charge the Mitchells for any of the time Bo spent working at their residence, took 10 percent off their bill as an apology, and did not charge for the extra time it took the two remaining members of his staff to complete the job. After picking up the company truck at Casey's Bar & Grill, Mr. Peterson returned to the Mitchell residence to apologize to Mrs. Mitchell for any inconvenience the actions of his former employee may have caused.

Mr. Peterson's business was predicated on high-quality work and exceptional customer service. That his firm had not advertised in well over 25 years and received nearly all of its business as a result of referrals from a long and dedicated list of customers spoke volumes as to the high professional standards by which he operated. Because Mrs. Mitchell had been Mr. Peterson's client for so long, she knew the lengths to which he would go to keep customers happy, and she had confidence the situation would be satisfactorily resolved.

Working Through Change: Developing as a Young Supervisor

OBJECTIVES

After studying this chapter, you should have an understanding of:

- adapting to change
- financial costs of change
- developing a workable game plan
- formulating a timeline
- amending the plan
- seasonality of the profession
- positive cash flow
- unpredictable changes

Change is everywhere, including in the Green Industry. Your job will be in constant flux as will be the jobs of those who you supervise. In fact, change in the horticulture profession is not measured on a day-to-day basis but rather by the hour and perhaps even by the minute. One of the hallmarks of successful and forward-looking managers is their ability to adapt to change.

Adapting to change enables entry and midlevel supervisors to deal with new and problematic situations. Change occurs frequently with today's rapidly developing technologies, and supervisors at all levels must adapt quickly to remain competitive. Often a young supervisor hardly has time to settle into his or her new position before a major crisis occurs. Perhaps another crew leader was hired but decided at the last minute to accept a position with another company. With spring right around the corner, you may now be assigned more responsibility with the firm. A recently hired manager can look on such a situation as a burden or as a chance to prove his or her worth to the company. Every young supervisor needs to learn to manage change as quickly and effectively as possible and to use sound decision making when working through new and challenging issues. When you feel engulfed in significant and often stressful change, evaluate all available options, consider their consequences, and then formulate a course of action. Embrace positive and meaningful change. By critically evaluating each new situation, managers can often redirect their efforts and work toward a successful solution. Don't forget to keep an eye on the financial costs of change.

When working through change, remember that

1. managers who adapt are more likely to advance quickly within an organization.
2. uncertainty is often opportunity in disguise.
3. firm decision-making is a necessary attribute to good management.
4. the short- and long-term goals of the organization should always be taken into consideration.
5. communication skills are extremely important.

Managing Organizational Change

To best manage change, consider the magnitude of the change required. Relatively minor changes are seldom worthy of a busy manager's time. On the other hand, catastrophic change often demands that a manager concentrate his or her efforts on solving significant issues. Supervisors should determine which problems they want to concentrate on and with what degree of urgency. They must then establish a suitable timeframe. Must the problem be dealt with by noon today, by the end of the month, or within 3–5 years? Managers must determine what strategies to use and implement them after developing a workable game plan. Developing a strategy or formulating a timeline is useless if no one is willing to take charge and lead; someone must have responsibility for overseeing the solution to each problem. Some may not be sympathetic to the changes being instituted. Watch for those inhibiting or even sabotaging the plan. Finally, amend each plan as it grows and develops. Even the best strategies will need to be altered as they progress toward the solution. Each of these methods is particularly valuable when managing problems associated with the seasonality of many horticultural businesses (Hrebiniak, 2005, pp. 228–229).

Few parts of the Green Industry are more seasonal than a large, wholesale grower or a commercial greenhouse operation. Examples are the many time-related issues that a wholesale grower needs to consider in delivering 50,000 field-grown chrysanthemums to vendors in time for the autumn sales season, or a commercial greenhouse operation's marketing of hundreds of thousands of annual bedding plants each spring. A director of horticulture at a large botanical garden needs to manage change to coordinate the fall planting of countless spring-flowering tulips and daffodils.

Dealing with Change

One of the difficulties in managing change is due to the seasonality of the profession. That difficulty can be viewed as a burden or a positive challenge. The number of months a company can operate profitable business activities shrinks dramatically the farther north it is. But utilities, insurance, mortgages, and other expenses demand payment throughout the entire year. Northern companies also need to maintain equipment and most likely retain a skeletal staff. In northern states such as Minnesota, Michigan, Vermont, New Hampshire, and Maine, precious little landscaping goes on in December, January, or February. In many areas, neither are November and March high-income months. Indeed, some landscape businesses earn supplementary income by selling Christmas trees or firewood. However, with Optimist Clubs, Rotarians, Boy Scouts, Girl Scouts, school groups, and church organizations all selling holiday trees, it is a competitive business. Others depend on the cash flow from commercial or residential snow-plowing contracts, which can provide positive cash flow at a financially difficult time of year. Snow plowing, however, has three main problems. First, plowing wet, heavy snow is hard on work vehicles. Second, pressure from low-ball competitors can seriously cut profit margins. Last, the hours are long and often grueling.

Supervisors in the Green Industry must remain adept at incorporating seasonal swings in their business cycle, such as the operation of a commercial Christmas tree farm, into their long-range plans. Just imagine the differences between May, June, and July and early November to late December. Business for commercial greenhouses is extremely cyclic as well. Florists have business peaks associated with Easter, Mother's Day, and Valentine's Day. The vast majority of poinsettias are marketed and sold in the few weeks leading up to Christmas. Although these cycles are for the most part very predictable, they nonetheless severely test the ability of supervisors in greenhouse and florist industries to adapt to daily if not hourly changes.

When initiating change, supervisors should

1. strive to encourage prior input from key staff.
2. think outside the box.
3. accurately evaluate the strengths and weaknesses of staff.
4. communicate the purpose of any change honestly and accurately.
5. stand behind potentially unpopular decisions and do not be apologetic.

Managing Change

Some changes can be planned for more than others. A florist can plan for and manage the sale of holiday flowers with reasonable predictability. Larval infestations of European chafer or Japanese beetle can be reasonably planned for by golf course superintendents. The unpredictable changes are the challenging ones. Suppose your most productive landscape crew leader was caught drinking at Wildwood Estates. Wildwood is a large and upscale seniors community that has been one of your most lucrative landscape maintenance contracts for the past 7 years. You terminated the crew leader's employment with your landscape company, and now you must replace him before arrival of the busy spring season, just around the corner. In the meantime, who will supervise your three landscape crews at Wildwood until a replacement can be hired and trained? How will you deal with Wildwood's management, who know that a police officer removed your landscape crew leader from Wildwood in a drunken state? What long-term damage control will you use to prevent losing this lucrative customer?

Obviously, not all change is as devastating. Still, supervisors at all levels need to plan for expected changes and try to anticipate others. Remember that changes must take into account the short- and long-term goals of the organization. For example, if your landscape crew leader's two-way company radio stopped working at the Ahmad residence this morning, how significant is that to the operation of the company? If the crew leader was not able to coordinate with the main office regarding a delivery of trees and shrubs, the radio system's failure affected the crew's productivity. Perhaps other communication-related systems are available and would better meet company needs. However, would the training period required to implement a more dependable and technologically advanced system in the midst of the current spring season itself affect productivity? In short, you look for the quickest solution that will allow the company to meet its productivity needs now and at the same time meet company needs in the future. It is crucial to understand how each piece of a business operation fits together not only in the short term but long into the future as well.

Evaluate each problem as it impacts the overall productivity of the operation. Be ruthless in identifying pitfalls that could develop into serious problems. It is not uncommon for one issue to interrelate with another, making a seemingly easy solution complicated. Moreover, for solving each problem, set realistic goals with respect to where the company is now and where it is planned to be in the future. Coordinate plans so that day-to-day operations progress smoothly and cash flow and other finances are not negatively affected.

When significant changes are made in an organization

1. make sure key staff have had sufficient input in the planning process.
2. be certain that new responsibilities have been clearly communicated.
3. make sure all relevant individuals support the new procedures.
4. don't underestimate the importance of education and training sessions.
5. understand that some staff dissatisfaction is to be expected.
6. make sure key individuals take part in the planning and implementation processes.
7. respect time frames and budgets.
8. honesty and integrity are vital components of its long-term and meaningful success.

Initiating Change

Try always to see change in as positive a light as possible. Young supervisors must consider change that benefits not only their own professional growth and development but also the long-term goals of the organization. Change allows supervisors to be a positive and proactive part of the planning process. Supervisors are more likely to initiate change if they choose their options with maturity and discretion. Change is a very important aspect of operating any progressive and forward-thinking company. It must be thought through carefully, as change with an eye to the future is the backbone of any well-run business. Be sure not to operate in a vacuum. Young managers should always consider the input of other managers and departments. Other supervisors will more likely support changes if their input has been taken into consideration. To avoid disruptions, broad, sweeping organizational changes should not be made all at the same time. Making changes over time is not always feasible, but it should always be given consideration. Do not initiate significant change for little or no reason. Meaningless and untimely change often causes subordinates to lose respect for management.

Building Trust

Supervisors need to consider how subordinates will react to newly introduced changes. Much depends on the type of change being initiated as well as its time frame. How change is communicated can often have a serious impact on how staff will react.

Some staff always resist any alteration to their routine, and there's not much that can be done about it. Such individuals are seldom respected by their peers. Moreover, they tend to view change only in light of how it affects them professionally and not as it relates to the overall good of the organization. Perhaps the best indicator of how staff will accept change is the level of trust they have for management. If staff believe that they have historically been treated with fairness and respect, they offer little resistance to significant, long-term change. However, if staff have been treated unfairly, they tend to distrust nearly any change that management initiates. Staff–management relationships can greatly affect a supervisor's long-term effectiveness and productivity.

CONCLUSION

To initiate change and manage staff-related resistance, first determine whether employee dissatisfaction is truly caused by the change. It behooves managers to investigate why staff may be unsupportive toward a seemingly positive change. Many times surprising reasons other than the current change underlie staff discontent. Sometimes the source of resistance is staff who do not seem to resist but who covertly sow negativity and distrust among their coworkers. Beware this type of employee, as he or she can poison staff morale. When supervisors meet active resistance to a proposed change they should not barter or make deals to buy subordinates' support. Some supervisors try to palliate particularly negative employees in an effort to win them over to their side. Most mature and seasoned supervisors would agree that such a strategy is at best misguided and is almost sure to fail.

Remember that clear communication is often key to the success of supervisors at any level. It is seldom advantageous for managers to apologize to staff regarding changes in company policy. By far the best approach is to honestly and professionally communicate with staff to reconcile differences they have with the change. Treat all staff complaints as important, as staff dissatisfaction may not always be based on accurate information. It is always best to consider staff concerns and strive to arrive at an amicable solution. Always keep in mind that supervisors will never be able to satisfy the concerns of each and every staff member or be successful in solving all problems. A negative reaction should not be the deciding factor for not implementing positive change.

Organizational change will be most successful if

1. there are measurable goals.
2. staff buy into the project.
3. the objectives have been clearly and adequately communicated.
4. a contingency plan is ready to be enacted.
5. staff are well aware of both the positive and the negative aspects of the plan.
6. all pertinent staff are sufficiently committed to the success of the plan.
7. the plan is malleable and can be refined over time.
8. cooperation and trust have been encouraged.
9. key staff believe their input has been listened to and appreciated.
10. the company has a history of introducing successful changes.

DISCUSSION QUESTIONS

1. What is the advantage for young supervisors who can quickly adapt to new situations in the workplace?
2. Give some examples of how a manager's promotability is reduced by being caught up in day-to-day activities and failing to plan for the future.
3. What are the benefits for young supervisors who have a positive outlook toward change?
4. Give some examples of the relationship between organizational efficiency and managing change.
5. The florist, greenhouse, and Christmas tree industries must be especially adaptable in the ways that they manage their businesses. What other Green Industry professions must be particularly adaptable to change?

6. As the owner of a landscape maintenance firm, how would you have dealt with the situation at Wildwood Estates as discussed in this chapter?
7. Was the situation at Wildwood Estates an immovable obstacle to the landscape maintenance firm's long-term relationship with its client?
8. How damaging do you think the situation involving the inebriated landscape crew leader at Wildwood Estates was to the landscape maintenance firm's relationship with other well-off clientele in the area?

   Do you think upscale clients communicate with each other about less-than-satisfactory business experiences when they chat at the country club, health spa, or indoor tennis facility?
9. Although it is often necessary to act early and decisively in managing change, does this give young supervisors the authority to circumvent company protocol or policy?
10. Why is it often best for significant organizational change to be instituted in stages?
11. Why is communication critical to implementation of change in an organization?

Geraldo Lara's Responsibilities as Assistant Horticulturist at Creekside Country Club

As one of three crew leaders at Creekside Country Club, Geraldo Lara had a wide range of horticultural responsibilities. Creekside was located in the Great Smoky Mountains region of Tennessee and had gained a solid reputation for the quality of its turf and landscape plantings. The Horticulture Department was responsible for the landscape plantings of all annual and perennial flowers as well as spring bulb displays throughout the course. Their primary responsibilities included the well-designed landscape areas that surrounded the clubhouse and adjoining areas. Horticultural display beds were also located along the 1st, 5th, 7th, 9th, 13th, 15th, 17th, and 18th holes. The responsibilities of each crew leader included caring for all trees and shrubs and horticultural maintenance.

Geraldo had been employed at Creekside for just over 4 years. He had been hired as a crew member after graduating with a baccalaureate degree in ornamental horticulture from a major southeastern university, and he had been promoted to crew leader 18 months ago. The two other crew leaders had 5 years or more of managerial experience, but the director of horticulture had confidence in Geraldo's abilities.

Geraldo and his crew were assigned the edging and mulching of the large landscape beds that surrounded the 18th green. William Rodgers, one of Geraldo's crew members, had been employed at the club for just over 2 years. William had not graduated from high school and had worked at three of the other clubs in the area in the past 6 years. Although William was technically proficient in his horticultural skills, he had a history of being hotheaded and not getting along with fellow crew members.

Geraldo instructed one crew member to pick up the necessary tools from the equipment-storage area. He sent William to water the perennials in one of the department's lathe structures, estimating it would take no more than 15–20 minutes, and told him to meet the rest of the crew at the landscape beds at the 18th green. Nearly an hour later, as Geraldo and his crew edged the beds, William finally approached the worksite. Geraldo began to question him as to why it had taken him almost an hour to water the perennials. William claimed the perennials were especially dry and that he was simply being a responsible employee by watering each pot as thoroughly as possible. Geraldo knew William was not being honest and that he had been shirking his responsibilities. William did not want to admit his laziness and the conversation quickly escalated into a heated discussion. After a few minutes, Geraldo finally told William to leave the property and expect to receive an official termination letter.

As William left the premises, he stopped by the maintenance shop to discuss the situation with Jack Armstrong, the chief union steward at Creekside. Mr. Armstrong was aware of William's temper and that the length of time to water the perennials had been excessive. But Geraldo had not handled the situation well and did not possess the authority to terminate a full-time employee. William, he knew, had been given oral warnings regarding repeated outbursts of anger, but no written documentation had been placed in his personnel file. After finishing his conversation with William, Mr. Armstrong stopped by the office of Janet Fujita, the club's director of horticulture. The two had developed a solid relationship in the 15 years they had been working through labor–management issues. After meeting with Mr. Armstrong for nearly 30 minutes, it was clear to the director of horticulture that William had not acted in a responsible manner and should be at least reprimanded.

Ms. Fujita also understood that her crew leader had not used sound judgment in handling William. Mr. Armstrong was correct that Geraldo did not have authority to terminate a full-time employee and that no letters of reprimand noting William's need for anger management had been placed in his personnel file. Given the circumstances, Ms. Fujita contacted William at his home and instructed him to report for work the following morning. She also called Geraldo and stressed the need for him to be much more careful in disciplining his full-time staff. Although Geraldo was not pleased with the outcome, he learned a valuable lesson in staff supervision.

QUESTIONS FOR DISCUSSION

1. Do you agree with Janet Fujita's decision to reinstate William Rodgers as a full-time employee at Creekside Country Club?
2. How do you think Geraldo Lara should have handled the laziness, lying, and anger-management issues involving William?
3. How do you think Geraldo should have handled the disciplinary process involving William once his anger-related issues began to arise?
4. Do you agree with the chief union steward's position that Geraldo overstepped the bounds of his authority in attempting to terminate William?
5. Do you agree with the chief union steward's position that letters of reprimand are almost always a critical component in documenting the work-related inadequacies of a full-time employee?
6. Do you believe that Janet Fujita made the right decision when she reversed the immediate termination of William?
7. What might have been the consequences to the club if Janet Fujita had not reinstated William as a full-time employee?
8. If you had been supervising William, how would you have disciplined him?
9. Do you believe that William's time as an employee at Creekside is limited?
10. If you were the director of horticulture at Creekside Country Club, how would you have handled your discussion with Geraldo?
11. If you were the director of horticulture, would you have chastised your young crew leader for allowing his discussion with William to become so heated?

CRITIQUE

William Rodgers was not a dependable and motivated employee of Creekside Country Club. Everyone was well aware of his shortcomings, including the chief union steward. It is important, however, to follow proper procedures when terminating a full-time employee. This is especially true in situations where management and union representatives must work closely together for the overall good of the organization. He deserved to be terminated, but Geraldo Lara used poor judgment in disciplining his crew member. Geraldo did not have the authority to terminate a full-time employee. The proper procedure was for Ms. Fujita to discuss potential terminations with the club's director of human resources. Such managerial foresight avoids costly and embarrassing lawsuits.

At the very least, Geraldo Lara should have discussed his dilemma with the director of horticulture and allowed her to handle the situation. A meeting with Ms. Fujita, Mr. Armstrong, William Rodgers, and the director of human resources would have probably resulted and a letter would have likely been placed in William's personnel file documenting his inappropriate behavior. As chief union steward, Mr. Armstrong would probably have suggested that William's continued employment at the club be contingent on his attending anger-management classes at the local community college. Even under the best of circumstances, William would have needed to seriously modify his behavior if he had any hope of maintaining his position as a full-time employee at the club. Although there are a few instances where it is permissible to take immediate action in the termination of a full-time employee, they are the exception rather than the rule.

Geraldo Lara was deficient in his managerial judgment for yet another important reason. Geraldo was unaware that his heated conversation with William Rodgers was carried on in front of Dr. Russell A. Smith, a longtime member of the club. Dr. Smith was finishing a particularly fine round of golf and missed a short 3 ft putt on the 18th green because of the confrontation. Bothered by the obvious lack of professionalism, Dr. Smith immediately contacted the general manager, who brought it to the attention of Ms. Fujita. Geraldo Lara should never have allowed a heated discussion to occur in such a visible and high-profile location.

Hiring Staff: Making the Right Choices

OBJECTIVES

After studying this chapter, you should have an understanding of:

- long-term goals
- job description
- compensation package
- promotability
- up-to-date résumé
- team-oriented approach
- affirmative action and equal opportunity employer
- basic understanding of Spanish
- special pesticide licensing commercial driver's license
- problem-solve difficult situations

Hiring the right employee in any profession is no easy task. Hiring qualified staff in the Green Industry can be especially difficult because of its seasonality. If you are hiring full-time staff, what personal and professional characteristics should you look for in a candidate and why are such decisions so important to the long-term growth of the organization?

Defining Your Needs

The hiring of full-time staff should always take into consideration the continued, long-term goals of the organization. The successful operation of any business depends on the quality of its employees. Thus it is important to ensure that each individual performs as a commited and responsible member of the team. It is not uncommon for those hired into low-level management positions to eventually prove themselves to be diligent and motivated employees. Such individuals often advance into middle management and from there into more demanding supervisory roles. Staff who rise quickly wthin an organization often display two primary characteristics. They are dedicated to the long-term goals of the organization and are focused and trustworthy employees. It is therefore worthwhile to evaluate candidates for even the lowest-level manager position with an eye toward the future.

Developing a Hiring Strategy

Before requesting applications, it is critical to evaluate employment needs. Not all vacated positions need to be filled. It may be more advantageous to allocate resources to other areas of the organization or reassign work-related responsibilities to other employees. If it is decided that the position needs to be filled, take the following steps to locate the best candidates. Provide a clear job description that accurately details the position's responsibilities. Be sure the position's technology requirements are consistent with the company's anticipated growth and candidates possess the necessary skills to keep pace with the changing needs of the industry. Offer a compensation package that is competitive within the industry. Select for interviews candidates whose level of education, work-related experience, and technical expertise meet or exceed the position's requirements.

When evaluating the suitability of prospective employees, a supervisor should closely evaluate the promotability of each candidate. This is especially true when striving to hire the best and most creative talent. In addition, for applicants being considered for management positions, note whether they present themselves professionally and produce an accurate and up-to-date résumé. A team-oriented approach to conducting important interviews is invaluable for evaluating potential candidates. It is imperative that young supervisors familiarize themselves with affirmative-action and equal-opportunity employer (AA/EOE) hiring practices.

When many candidates are competing for the same position look for differences that may set one above the rest, such as being reasonably fluent in Spanish or having special pesticide licensing as well as a commercial driver's license. Try to discern a candidate's level of motivation, honesty, and ability to adapt to difficult situations.

Advertising the Position

The methods used to advertise a full-time position should be in keeping with the type and level of job being offered. A shotgun approach can result in large numbers of responses that are cumbersome and time-consuming to manage. Moreover, the majority of responses from nontargeted advertising will seldom be a good match for most positions. A focused approach is almost always the most beneficial for attracting dependable and motivated staff.

Advertisements placed in trade-related newsletters and magazines reach well-qualified candidates. Depending on the type and level of position offered, horticulturally-oriented employment agencies and Internet Web sites are resources to consider using. Dollars spent in locating qualified and dependable staff is money well spent.

Selecting Candidates to Interview

Analyze each résumé and cover letter for accuracy. Does the résumé obviously falsify important information? Does the date that the applicant graduated from college mesh with his or her overall employment history? Is the individual inflating his or her professional credentials? Are the cover letter and résumé visually attractive, grammatically and syntactically correct? If an individual fails to produce an accurate and presentable résumé, his or her landscape-installation proposals are unlikely to be any better. A committee should be appointed to assure objectivity throughout the entire interview procedure. Once the letters of application and résumés have been adequately evaluated and candidates selected to interview, be sure to send a prompt and courteous letter of rejection to each applicant not selected.

The Interview

The next step is to schedule on-site interviews. Interviews may need to be scheduled outside normal business hours. Set aside sufficient time for interviews: at least an hour for the actual interview, additional time if a tour of the facilities is to be included, and enough time for the interview team to discuss the pros and cons of each candidate after an interview. Provide an interview environment that is conducive to a positive exchange of information; each interviewee should feel comfortable and not intimidated. Prevent interruptions during the interview by holding telephone calls and minimizing disruption from fax machines, copiers, and computers. Although it is important to ask job-related questions, it is equally valuable to allow time to simply get to know the interviewee. Much can be learned from seemingly idle conversation at the start of an interview session. If, for example, a candidate is interviewing for a sales-related position and has difficulty carrying on a simple conversation, he or she has little hope of succeeding in the position. An arrogant candidate who cannot stop talking about himself or herself may not be the right choice either. At some stage, allow the candidate the opportunity to ask questions about the position and the long-term goals of the organization.

The interview team should be prepared to answer questions about the compensation package associated with the position. Do not take the chance of losing a highly qualified candidate by not being able to answer rudimentary questions regarding retirement, health insurance, and vacation benefits.

CONCLUSION

During an interview session, probe the applicant's ability to problem solve difficult situations. Ask questions that encourage thoughtful answers. For instance, ask interviewees what they believe their greatest professional strength is, and follow up with a question about their most significant weakness. Questions that allow interviewees to talk about their long-term career goals can help evaluate their suitability for full-time employment. There will be times when the professional capabilities of an applicant will clearly make him or her the most qualified individual for the position. However, most candidates have weaknesses along with their positive characteristics. The key is to seek out the applicant with the highest potential of being successful in the position. Do not prejudice yourself either for or against a particular candidate. Some applicants will have glaring and undeniable deficiencies in their professional qualifications; others will have stellar qualifications. Remain objective and always use good judgment in selecting the right person for the job. A number of years ago I was asked to participate in the selection process for the director of marketing of a nationally prominent nonprofit institution. The senior vice president overseeing each interview was obviously sold on the professional abilities of a particular candidate. However, the candidate had held no less than five full-time positions in the last 4 years and was unable to give a reasonable explanation for changing jobs so frequently. Because the position demanded a reasonable commitment toward the long-term goals of the organization, the individual was clearly ill-suited for the position. As it worked out, the candidate was hired but left the position within a few short months.

Solicit the opinions from other trusted staff regarding each candidate. Some applicants try to act their way through an interview but show their true personalities to those they believe have no direct input in the hiring process. If applicants wait in a lobby, how do they interact with the nearby receptionist? Cordial and polite or overly nervous? Confident or boastful and arrogant? If candidates tour the facility with someone who was not at the interview, are they respectful or standoffish and rude? Do not rush to select; wait until each qualified candidate has been interviewed. No matter how critical it is to fill a vacant position, never hire an individual who you believe is not well-suited to the responsibilities of the position. Finally, remember that there will never be a perfect candidate. Some candidates may not have quite the educational background that you had hoped for or may not be quite as strong in their technical expertise or job-related experience. Perfect candidates simply do not exist, and there will always be some compromise in filling a job or position in any organization. Make your final decision by weighing the pros and cons of each candidate. Hiring qualified staff is never an easy task, and managers and supervisors must work to assure that a quality individual is chosen.

DISCUSSION QUESTIONS

1. Why would the seasonality of much of the horticulture profession make hiring qualified staff difficult?

2. Why is it important to coordinate employment needs with the long-term goals of the institution?

   What relation does this coordination have to an organization's success in retaining top-quality staff?
3. It is imperative that job descriptions be carefully written. Find some Green Industry job descriptions and critique them. What could be changed to improve them?
4. Do most horticulture-related organizations have a director of human resources on staff? Why is such a position so important to the long-term success of a business?
5. Give three reasons why the opportunity for promotion in an organization is of prime importance to retaining capable and motivated staff.
6. What does AA/EOE stand for?

   Can you be an efficient and effective manager in the Green Industry and not have a working knowledge of AA/EOE standards?
7. How are a candidate's technical knowledge, formal education, and professional experience linked to each other?

   Is it possible for lack of formal education to be more than compensated for by an individual's professional experience within the Green Industry?
8. Why consider the hiring process as a coordinated effort composed of many interrelated parts?

   Explain in detail what you see as the most important component of the hiring process.
9. Why is a team-oriented interview so important to successfully choosing the right candidate for a key position?
10. Would you use the information presented in this chapter as a guide for employment opportunities that you might apply for in the future?

    Would you be able to hone your own personal interviewing skills using this information?

Timothy Wilkinson's Hiring as the Assistant Production Manager at Spring Grove Nursery

Timothy Wilkinson was completing the spring semester of his senior year at a well-known university in the southeast. Timothy had attained a 3.57 GPA throughout his 4 years in college and had worked diligently to maintain a 4.0 GPA in all of his horticulture courses. During the summers of his sophomore and junior years, Timothy had worked with Carlson Brothers' Landscape & Maintenance. Carlson Brothers' had served southern Connecticut for over 25 years. Timothy enjoyed working at Carlson Brothers' and had gained a great deal of practical, hands-on experience.

Timothy learned the duties of his position quickly and by the end of his second summer had been given significantly increased responsibilities. Although the firm's landscape architects were the primary contact with each client, Timothy had developed superb customer-service skills and was looked on by senior management as an exceptional team player. As Timothy was leaving at the end of his second summer, Carlson Brothers' encouraged him to consider a full-time position as a customer-service representative for them after graduation. Customer-service representatives assisted the firm's landscape architects in solving client-related problems. Although Timothy had very much enjoyed working at Carlson Brothers', he was not sure whether full-time employment within the landscape industry meshed with his long-range professional goals.

Early in the fall semester of his senior year, Timothy had begun to focus on the wholesale nursery industry as his career choice. His interest was largely fostered by Morris, a horticulture classmate and close friend. Morris's uncle ran a highly respected, 165 acre nursery that specialized in field-grown B&B trees and shrubs. Morris and Timothy worked at the uncle's nursery in the fall and spring digging seasons and gained valuable hands-on knowledge within the trade.

One of the professors at the university told Timothy that a well-respected 425 acre B&B nursery across the state was interested in hiring an assistant production manager. After carefully reading the job description for the position, Timothy noticed that the nursery offered a good starting salary, promised expanded responsibilities for the right candidate, and had a profit-sharing plan. It was in the process of acquiring an additional 145 acres to expand its inventory of shade and small flowering trees. The nursery was only 2 hours away from the university, which would allow him to visit his girlfriend, Xiaojian, on weekends. Interested applicants were directed to e-mail, fax, or mail their résumé to Mr. Wilson Douglas, the co-owner and general manager of Spring Grove Nursery. That weekend, Timothy and Xiaojian looked at the nursery's Web site and saw that the nursery offered a wide selection of ornamental trees and shrubs of the highest quality. The nursery was well established and was now in its third generation of ownership. It was founded over 65 years ago by Wilson Douglas's grandfather. They agreed that the job warranted further investigation.

Timothy and Xiaojian fine-tuned his cover letter and résumé and e-mailed them to Mr. Douglas. Little more than a week later, Timothy received a call from Mr. Douglas's secretary to set up a telephone interview. They agreed on the following Thursday at 4:00 p.m. The telephone interview went well.

Mr. Douglas asked numerous questions regarding Timothy's course work and work-related experience, and he encouraged Timothy to ask about Spring Grove Nursery. They arranged to meet the following Saturday at 9:00 a.m. for an on-site tour of the facility.

As Timothy drove up the road leading to the nursery, he noticed a man throwing sticks for two Labrador retrievers on a home's front lawn. The man was Mr. Douglas, whose home was directly adjacent the nursery. Mr. Douglas waved to Timothy and walked out to greet him. Timothy offered Mr. Douglas a firm handshake, and after some light conversation, Mr. Douglas and Timothy set out on a tour in Mr. Douglas's pickup truck. Mr. Douglas ribbed Timothy about wearing a three-piece suit but quickly put the young man at ease by complimenting him on his appearance and approvingly noting that Timothy was nearly 15 minutes early for their meeting. As they drove through the nursery, Mr. Douglas quizzed Timothy on the identification and culture of row after row of trees and shrubs. Timothy, silently grateful for the two semesters of his senior year spent as a teaching assistant for woody-plant identification courses, impressed Mr. Douglas with his general knowledge of woody ornamental plants, although he was unsure of some of the more esoteric cultivars.

After walking through a few of the newly planted plots and touring the equipment barns, Mr. Douglas and Timothy walked back to the office to talk more specifically about the position. As they entered the double-wide trailer, Mr. Douglas introduced Timothy to the assistant office manager and his wife, who oversaw all business-related aspects of the nursery. Both of them were working this Saturday morning because of the nursery's heavy volume of spring orders. As Timothy entered Mr. Douglas's office, he couldn't help but notice the numerous plaques and citations that almost covered three full walls of the room. Some were from the North Carolina Nursery Association and commended him for his service as both a past president and a former member of the board of directors. Other plaques recognized Mr. Douglas for his membership and support of service organizations throughout the region. One large plaque specifically commended Mr. Douglas for leading the fund-raising efforts of a local hospital's $35,000,000 capital campaign. Timothy also noticed that Mr. Douglas had earned his baccalaureate and master's degrees in ornamental horticulture from the same university Timothy attended.

After approximately 20 minutes, Mr. Douglas asked Mrs. Douglas to join the discussion. The three talked for an additional 30 minutes, when Mrs. Douglas excused herself by citing the amount of work awaiting her. As she left, she told Timothy how genuinely impressed she was with him and that she appreciated his coming to meet with her and Mr. Douglas. After his wife left, Mr. Douglas asked whether Timothy had any additional questions regarding the position of assistant production manager at the nursery. Timothy said he had no further questions but he wanted to say that he was impressed with the operation and very interested in the position.

In reply, Mr. Douglas offered the position to Timothy on the spot. He requested that Timothy accept or reject the position within 2 weeks and told him he could ask either himself or Mrs. Douglas any additional questions that occurred to him. In Wednesday afternoon's mail, Timothy received a letter from Mr. Douglas's secretary containing a detailed summary of the compensation package. After discussion with Xiaojian, Timothy decided to accept the position and called Mr. Douglas Friday morning. Timothy worked for 8 very productive years with the Douglases before starting a wholesale nursery of his own. He had saved considerable capital during his employment at Spring Grove and was able to purchase 177 fertile acres in a neighboring county.

QUESTIONS FOR DISCUSSION

1. Timothy had attained a cumulative 3.57 GPA during his 4 years in college. How much emphasis do most prospective employers place on a student's academic record?
2. How valuable was Timothy's employment experience with Carlson Brothers' Landscape & Maintenance in preparing him for a career in the wholesale nursery industry?
3. How well suited do you believe Timothy would have been for the position of customer-service representative with Carlson Brothers' Landscape & Maintenance?
4. How valuable was it for Timothy to have gained additional experience in the horticulture trade by working at the nursery owned by his friend Morris's uncle?
5. If you were Mr. Douglas, would you have conducted an initial telephone interview with Timothy before inviting him for a tour of the nursery?
6. What is your opinion of Timothy wearing a three-piece suit to an interview at Spring Grove Nursery?

   What is appropriate attire for an on-site interview at a wholesale B&B nursery?
7. During an interview session, would you expect to be given a tour of a B&B nursery in a pickup truck? Why or why not?
8. Overall, how do you think Timothy conducted himself during his interview session at Spring Grove Nursery?

   What could Timothy have done differently to further support his chances of being offered the position of assistant production manager by Mr. Douglas?
9. How integral was Mrs. Douglas to the selection of Timothy for the position of assistant production manager at Spring Grove Nursery?

Hint: analyze and evaluate the entire interview scenario before answering this question.

10. How much do you think the character and personal integrity of Mr. and Mrs. Douglas influenced Timothy in his decision to accept the position at Spring Grove Nursery?
11. How much influence do you think Mr. and Mrs. Douglas had on the future success of Timothy Wilkinson as a wholesale nursery owner?

CRITIQUE

The interview at Spring Grove Nursery between Mr. Douglas and Timothy Wilkinson did not follow accepted protocol. But it was a success for Timothy and Mr. and Mrs. Douglas. The point is that graduating seniors must be perceptive and anticipate the type of interview they most likely will be involved in. A college senior majoring in horticulture should know to expect part of the interview at a wholesale B&B nursery to be held bouncing along dirt roads in a pickup truck.

That Mr. Douglas had attended the same university Timothy was attending was a good indication that Mr. Douglas would contact professors he knew there to see what they had to say about Timothy. For his part, Timothy probably could have done a more thorough job of investigating Wilson Douglas and Spring Grove Nursery beyond simply looking at the operation's Web site. Mr. Douglas set up a telephone interview with Timothy before inviting him to the nursery for a personal interview because he was in the midst of a busy spring digging season. He did not have a moment to spare. If Spring Grove Nursery fell behind in its digging schedule, the Douglases might not have been able to make up the financial loss in the remainder of the year. That is how critical the spring season is to nurseries whose lifeblood is digging field-grown shade and flowering trees. For Mr. Douglas to have set aside an entire Saturday morning to interview Timothy at this time of year speaks volumes about his commitment to choosing the right person for the job.

Regarding Mrs. Douglas and her role at Spring Grove Nursery, neither you nor Timothy knew that she had earned a master's in business administration and had completed additional graduate-level course work in accounting. She had full oversight over the financial health of the nursery, was an astute entrepreneur, and directed the entire business side of a sizable and extremely productive nursery operation. Don't be misled by the relatively short time that she participated in the interview. Mrs. Douglas had of course given Timothy's résumé close scrutiny and had discussed Timothy's qualifications with Mr. Douglas before his arrival at the nursery.

Although Timothy would be in the operations side of the nursery and Mrs. Douglas was in the office, if she had not been impressed with him, Timothy would not have been hired. The overall management of Spring Grove Nursery was very much a joint effort between the Douglases. When Mrs. Douglas told Timothy she was genuinely impressed with him and sincerely appreciated his coming to meet with them that morning, she was indicating to her husband that she liked what she saw in Timothy and that she approved hiring him as their new assistant production manager.

Evaluating Staff Performance: Overseeing Others in the Workplace

OBJECTIVES

After studying this chapter, you should have an understanding of:

- performance appraisals
- measurable and objective goals
- self-assessment of accomplishments
- regularly scheduled and formal discussion
- employment compensation
- written documentation
- employee productivity
- employee job satisfaction
- staff's strengths and weaknesses

All supervisors have the responsibility of guiding their staff to be more productive. An employee evaluation is an important part of this responsibility. Performance appraisals should focus on the positive growth and development of the employee. The Green Industry is a melting pot of cultures and ethnicities, but race, age, or religion have no part in any evaluation. It is imperative to always remain objective in any staff evaluation.

Evaluate an employee only for those situations that he or she has control over. If your crew leader parked the company pickup truck in front of a client's home and a hit-and-run driver sideswiped the vehicle, the employee should not be penalized for the carelessness of another driver. Focus on measurable and objective goals in a performance review. Staff evaluations should not contain surprises. If problems have developed since an employee's last review, they should have been addressed at the time they occurred. It is often valuable for staff to complete a self-assessment of accomplishments since their last review. This provides an honest evaluation of their work. It also allows you, their supervisor, to understand the employee's perception of his or her responsibilities.

Staff evaluations are an excellent opportunity for two-way communication between supervisor and employee with regard to an employee's work-related performance. The supervisor should never dominate a review session. Productive reviews require an honest discussion. If the organization is large enough to have a human resources specialist, supervisors should work closely with the specialist in completing staff evaluations. This is especially true when evaluating a marginal employee, especially one recently placed on probation. ❧

Elements of Evaluations

A performance review should be a regularly scheduled and formal discussion. The supervisor and employee should jointly review the subordinate's work-related performance. Notice that communication is central to a review. The appraisal process should include a sharing of thoughts regarding the strengths and weaknesses of the employee. It should also include strategies to help improve the employee's future performance as a productive member of the organization. The review should be documented in writing and include employee-related goals. Young supervisors should use performance appraisals as a learning tool for their employees. This means spending more than 15 minutes on an appraisal. An adequate review of strengths and weaknesses demands far more time and effort. In many ways, a performance evaluation should reflect how forward-thinking and proactive supervisors manage on a daily basis.

Many firms within the Green Industry tend to use generic evaluation forms to appraise employees. Generic evaluations encourage consistency throughout all levels of the organization, but they are designed to be used by almost any industry—plumbing companies, restaurants, telemarketing firms, amusement parks. As a result, generic a form cannot address the performance requirements specific to the Green Industry. A performance review should evaluate an employee's productivity. This includes not only the quantity of his or her work but the quality as well. The employee's efficiency should likewise correspond to his or her role within the company. For example, if the employee is in sales, consider proficiency in customer service. Performance reviews ensure that employee skills continually develop. Reviews look at how well an employee's skills match the position's requirements, how adaptable the employee is when upgrading skills, how well the employee learns more advanced technology, and whether or not additional job training is necessary for an upcoming promotion. Training should also encourage employees to remain focused on the long-term goals of the organization (Figure 14-1).

Employee job satisfaction is another element of a performance review. Is the employee dissatisfied with his or her current role within the company, or has he or she recently been passed over for a promotion? Is the employee unhappy with new responsibilities or insecure about his or her future within the organization? A review determines whether the employee is professionally challenged by his or her role within the company. There are obviously a myriad of questions to ask about employee satisfaction. Answering them is time well spent in promoting a positive and proactive work environment.

To most fully evaluate staff performance, consider

1. how close the employee came to achieving his or her goals.
2. the positive or negative influence the employee has had on the morale of others within the department.
3. how the employee's work-related performance has measurably improved since the last review.
4. the working relationship between the employee and coworkers.
5. the respect the employee shows for company property.
6. the quality and attention to detail the employee shows in completing assignments.
7. the overall productivity of the employee.

Continued

EMPLOYEE EVALUATION

Employee's Name Position/Title Date of Review

Department Supervisor Date Employed

Type of Review: ____________ Probationary ____________ Six Month ____________ Annual

Date of Last Evaluation ____________

Total days absent ____________ Total days late ____________

Quantity of Work:

_______ Unacceptable _______ Fair _____ Satisfactory _____ Very good _____ Excellent

Comments:

Supervisor's Signature

Volume and consistency

Quality of Work:

_______ Unacceptable _______ Fair _____ Satisfactory _____ Very good _____ Excellent

Comments:

Supervisor's Signature

Accuracy and neatness

Initiative:

_______ Unacceptable _______ Fair _____ Satisfactory _____ Very good _____ Excellent

Comments:

Supervisor's Signature

Motivation

Judgment:

_______ Unacceptable _______ Fair _____ Satisfactory _____ Very good _____ Excellent

Comments:

Supervisor's Signature

Problem-solving abilities

(continued)

FIGURE 14-1 Employee evaluation form.

Planning work, making decisions

Adaptability:

_______ Unacceptable _______ Fair _______ Satisfactory _______ Very good _______ Excellent

Comments:

____
Supervisor's Signature

Adjusts to change

Cooperation:

_______ Unacceptable _______ Fair _______ Satisfactory _______ Very good _______ Excellent

Comments:

____
Supervisor's Signature

Getting along with others

Speed:

_______ Unacceptable _______ Fair _______ Satisfactory _______ Very good _______ Excellent

Comments:

____
Supervisor's Signature

Rate of work

Job Knowledge:

_______ Unacceptable _______ Fair _______ Satisfactory _______ Very good _______ Excellent

Comments:

____
Supervisor's Signature

Technical proficiency

Since last evaluation, employee has

_______ Improved _______ No change _______ Regressed

Recommended for pay increase

_______Yes _______ No

Overall impression of this employee:

_______ Unacceptable _______ Fair _______ Satisfactory _______ Very good _______ Excellent

Comments:

____
Supervisor's Signature

FIGURE 14-1, continued

Continued

8. the overall dependability of the employee.
9. the employee's level of initiative or motivation.
10. the employee's punctuality in completing assigned tasks.
11. the ability of the employee to adhere to company policies and procedures.
12. the ability of the employee to work toward the long-term goals of the organization.

To maximize staff performance,

1. encourage employees to make positive and well-thought-through decisions.
2. make sure that staff have a clear understanding of the organization's goals.
3. allow employees sufficient individuality and creativity.

Continued

The final category which is often discussed as part of a performance review is compensation. Although a staff member's financial compensation should be evaluated by quantifiable means, this can be problematic within the horticulture profession. If a worker makes parts in a factory, a supervisor can easily count the number made. If a worker picks oranges in Florida or grapes in California, the same holds true. In the horticulture profession many jobs are difficult to quantify. In addition, the seasonality of the trade makes monitoring an employee's progress that much more difficult.

A 2006 survey of employee compensation outlines current trends in the Green Industry (2006 Wage & Benefits, 2007). Approximately 88 percent anticipated wage increases for staff, compared with 71 percent within the previous year. Companies specializing in landscape design, installation, or maintenance accounted for 54 percent of industry revenue, a significant increase from the 28 percent of the previous year. However, revenue earned by wholesale nursery growers, greenhouse growers, and retail garden centers in 2006 decreased significantly from 2005.

These data tell us that if companies have higher earnings, employees are likely to see increased wages. Likewise, depressed earnings result in lower staff compensation the following year. In other areas the survey found that the number of companies offering 401k retirement plans to salaried personnel had risen the last 3 years in a row. In 2004 only slightly more than 25 percent of salaried employees were offered 401k plans, whereas 2005 saw an increase of almost 6 percentage points, to 31 percent. Salaried staff with 401k plans increased again in 2006, to an impressive 42.9 percent. Tuition reimbursement likewise continued to rise. By 2006 nearly 36 percent of managers and supervisors were eligible for tuition reimbursement. Paid vacations and holidays were available to 76 percent of managers and supervisors, and 57 percent were eligible for an annual bonus. A total of 33 percent were allotted profit sharing, and 38 percent were eligible for family health care. Slightly over 26 percent had life insurance, and 19 percent were included in a company dental plan. Finally, nearly 10 percent of managers and supervisors had eye care and 31 percent were involved in a pension plan. Companies provided either uniforms or a uniform allowance to 45 percent, and nearly 60 percent received retail or services discounts (2006 Wage & Benefits, 2007).

Fine-tuning Performance Appraisals

As a supervisor, it is important to accurately critique an employee's strengths and weaknesses, give praise, and offer solutions. Has there been an increase in Clare's motivation and overall level of performance since her last review? What is responsible for William's recent upsurge in productivity? What is the best way to encourage Adriana to maximize her time-management skills? What is a fair and adequate measure of Rajiv's technical skills since his last review? Should Philip's next review be rescheduled for an earlier date because of his continued inability to perform at an acceptable level? How can Betsy be best rewarded for her consistently high levels of achievement? Although staff should motivate themselves toward increased productivity, their supervisor must always assume the responsibility of guiding them in their professional development.

Continued

4. encourage staff to fully support other members of the team.

Evaluating Problem Employees

It is important to curtail negative employee behavior, primarily to prevent them from influencing other members of the team. Solving significant personnel-related issues strengthens the company's competitiveness and improves its financial health, which ultimately depends on the commitment of competent and motivated employees. Whether working in a profit-oriented landscape company or a nonprofit botanical garden, all managers need to be proactive in strengthening the long-term growth of the organization.

CONCLUSION

The development of productive and forward-thinking employees should be the guiding philosophy behind any performance evaluation. Goals and objectives of businesses may vary, but the future of any organization is dim without a dedicated team of quality employees. Managers must quantify the strengths and weaknesses of their staff and use performance appraisals to support the long-range goals of the organization. Retaining the best and most qualified staff positively affects the future development of any organization. Recognition awards, salary-based compensation, and even promotions often hinge on the strength of an employee's performance evaluation. From an even broader perspective, an individual's professional goals are often influenced by his or her performance evaluations.

Performance reviews evaluate measured progress of employees and set realistic goals with input from both management and staff. Two-way communication is a critical part of the process. Always establish a clearly defined process to accurately evaluate the strengths and weaknesses of staff.

When evaluating the overall performance of employee teams, make sure that

1. each member clearly understands his or her role within the group.
2. the personalities in the group are reasonably compatible.
3. individuals acting in a leadership capacity work in cooperation with others in the group.
4. the group supports the strengths and weaknesses of each member.
5. each member supports the best interests of the group.
6. all members are able to accept constructive criticism.

DISCUSSION QUESTIONS

1. Why is an honest critique of an employee's strengths and weaknesses the most effective way to nurture staff as productive components of an organization?
2. Why should race, age, and religion play no part in an employee's performance evaluation?
3. Why is it important for an employee to complete a self-assessment of his or her professional abilities?

Before initiating an employee incentive program

1. understand that financial incentives are frequently viewed as short-term motivators.
2. realize that a bonus system does not always have to be financially based.
3. consider awards that are personalized to each recipient.
4. understand that simple means of rewarding employees can be very effective.

4. What does two-way communication, between an employee and his or her supervisor, contribute to a proactive and successful review?
5. Why should the goals of staff be coordinated with the long-term goals of the organization? Why should this be covered in a performance review?
6. Should employee satisfaction be a component of a performance review?

   Why is it important to employee retention and productivity?
7. Outline specific steps you would take to quantify an employee's productivity within the horticulture profession?
8. What is the connection between improving staff productivity and an organization's financial growth?
9. How do performance appraisals affect an organization's competitive status?

   How does a company's competitive status affect its cash flow and ultimate financial health?
10. How should performance appraisals be used in the planning and development of staff training?
11. Why should management be role models and be accountable for their actions?
12. How can an organization best establish specific and measurable goals for evaluating an employee's accomplishments?
13. How are hiring procedures, performance appraisals, and long-term financial growth interrelated with the sound management of an organization?
14. Do you see a positive and proactive performance appraisal system as being an integral managerial component of a company that you may own in the foreseeable future?

Felicity Keiffer's Performance Appraisal with Maria Chavez Design & Landscape

Felicity Keiffer graduated with a degree in landscape architecture from a large midwestern university. She accepted a position as landscape architect with Maria Chavez Design & Landscape and relocated to Boston to accept the position, where she enrolled at a nearby university and earned a second degree in horticultural science. In her 3 years with the company, Felicity had grown tremendously. The company was well known throughout the Boston area for designing, installing, and maintaining the highest-quality landscapes and had an equally well-deserved reputation for its quality-oriented customer service. Working directly with two of the firm's crew leaders, Felicity increased crew members' production and decreased their tardiness and absenteeism. Staff motivation increased in part because of the fairness and respect Felicity showed crew members. She had an engaging personality, which charmed clients who met and interacted with her.

As December approached Ms. Angela Guzman, the firm's vice president of design and installation services, scheduled an appointment with each of her landscape architects to review their progress and performance throughout the previous year. Ms. Guzman scheduled a 1.5 hour meeting with Felicity and invited Elizabeth Anne Shapiro to participate. Ms. Shapiro was in her 27th year with the firm as a registered landscape architect. Ms. Guzman mentioned to Felicity before the meeting that their time together would be used to review her performance and establish further development opportunities for her within the company.

At the meeting, Ms. Guzman, Ms. Shapiro, and Felicity discussed the young supervisor's level of accomplishment since her last review. Ms. Guzman said Felicity had developed an excellent working relationship with other landscape architects who she had coordinated with on various projects. Felicity had likewise gained the respect of the workers on the two crews that she supervised. Ms. Guzman commended Felicity for her timely and accurate completion of managerial paperwork. Ms. Guzman also complimented Felicity on her ability to motivate her crews. Because Felicity had not missed a single day of work since her last review, Ms. Guzman praised her for her reliability and told her she was an exemplary, positive, and professional role model to her crews. Ms. Guzman noted that Felicity worked well with minimal supervision and had showed commendable day-to-day judgment. However, she added, on a few occasions Felicity might have benefited from coming to her for advice in problem solving specific situations. She remarked that Felicity's long-term vision meshed extremely well with the long-term goals of the company.

The 1st hour of the performance review was devoted to Felicity and Ms. Guzman discussing Felicity's professionalism and productivity. Felicity had progressed well within the organization's ranks, and Ms. Guzman had ideas for Felicity's managerial advancement within the firm. One of the options that Ms. Guzman and Ms. Shapiro had discussed was for Felicity to become Ms. Shapiro's assistant. As the firm's senior landscape architect, Ms. Shapiro oversaw six full-time landscape architects, each responsible for mentoring a summer college intern majoring in landscape architecture. Felicity had proved herself to be dynamic and highly talented, and Ms. Guzman believed that both she and Ms. Shapiro would benefit professionally. Working side by side with Ms. Shapiro would allow Felicity to increase her managerial experience and her hands-on landscape architecture skills. Felicity would grow immeasurably from working under her direction. Ms. Shapiro was a highly regarded landscape architect, with numerous award-winning designs to her credit. Felicity greatly appreciated the opportunity of working with her. Although Ms. Shapiro had a reputation as being very demanding to work for, she was likewise known as an honest and forthright mentor who was dedicated to sharing her knowledge and nurturing the next generation of landscape architects.

Although the meeting lasted over 2 full hours, it was extremely productive for all. Ms. Guzman had the written summary of Felicity's performance appraisal as well as their ensuing discussion on advancement opportunities completed and placed in Felicity's personnel file. Felicity started as Ms. Shapiro's full-time assistant in mid-January and eventually started her own extremely successful landscape architecture firm.

QUESTIONS FOR DISCUSSION

1. Describe four of Felicity's primary professional attributes that helped contribute to her success.
2. Felicity had worked diligently to increase her crew's performance and overall professionalism. Do you think that was part of her job description or were Felicity's actions self-initiated?

   What other areas of expertise could Felicity have concentrated on to further improve the quality of her crews? Be specific in answering this question.
3. Do you think Ms. Guzman's comment regarding Felicity's occasional need to seek advice was justified, or do you think she made it in an effort not to award an employee with a perfect review?
4. Why do some managers avoid giving even their most accomplished staff a perfect performance appraisal? Would you give an employee such as Felicity a perfect performance appraisal? Fully support your thoughts on this managerial philosophy.
5. What additional attributes will Felicity need to develop to better prepare for higher-level managerial responsibilities? Be specific in your answer.
6. How important was it for Felicity's professional vision to coordinate with the long-term goals and objectives of the organization? Fully support your answer to this question.
7. How fortunate was Felicity to have gained the opportunity to work under Ms. Shapiro's tutelage?

   Would you accept an assistant's position under the direction of an individual such as Ms. Shapiro?
8. Although Ms. Shapiro was professionally demanding, she was known as a fair and supportive supervisor. Do you believe you would learn and grow as a landscape architect under such a mentor?

   Be specific in listing the positives and negatives of being an assistant to such an acclaimed professional as Ms. Shapiro.

9. How many years should assistants be mentored by an individual such as Ms. Shapiro before breaking away and doing their own mentoring of a young and soon-to-be accomplished professional? Fully support your opinion.
10. Why should an assistant move along and nurture another individual new to the profession?

CRITIQUE

Working hard and excelling, becoming known as a highly motivated supervisor, and becoming a respected employee with Maria Chavez Design & Landscape was in Felicity's best interest. Young managers can never be sure how their career will unfold. They never know how mentors and other professional contacts will benefit them in reaching their future career goals. As a result of Felicity's maturity, motivation, and dedication, she was rewarded with a very special opportunity early in her career.

Let us evaluate the importance of Felicity's association with Ms. Shapiro from Ms. Shapiro's perspective. As mentioned, Ms. Shapiro was a highly acclaimed landscape architect with numerous professional accolades. She would have had little difficulty convincing the best and most highly talented assistants to work with her. Why did she choose Felicity as her assistant? Ms. Shapiro was looking for an individual with a high degree of technical expertise, a keen intellect, and an amiable personality.

Highly regarded professionals do not want to waste time nurturing an individual who is not motivated and not committed to his or her chosen avocation. Felicity was an intelligent and forward-looking team player, interacted well with peers and subordinates, supported the firm's long-range goals, and was a respected professional who could be depended on to follow a focused career path. Felicity had not only been a proactive and dedicated employee but was also an industrious and hard-working student. In addition, Ms. Shapiro detected in Felicity a heartfelt dedication to succeed. It was most likely Felicity's personal commitment to work at her professional best that most convinced Ms. Shapiro to choose Felicity as her assistant. Although a résumé cannot capture such attributes, it was fortunate for Felicity that Ms. Shapiro saw firsthand the high level of her dedication and motivation.

Dr. Daniel Wilson's chapter entitled "Performance Appraisal," in *Effective* HR *Measurement Techniques*, has sample job descriptions and performance-appraisal worksheets. The author highly recommend the readers to become familiar with Dr. Wilson's chapter, for future careers within the Green Industry.

Forging a Career Path: How to Succeed in the Long-Term

OBJECTIVES

After studying this chapter, you should have an understanding of:

- quality internship experiences
- hands-on experience
- technical skills
- managing work-related stress
- organizing the work space
- personality traits
- bargaining process
- crisis situations
- negotiating skills

The time to start forging a career is well before accepting your first full-time position. Many argue that it begins as a freshman or sophomore in college, when students participate in quality internship experiences. Students enrolled in baccalaureate horticulture programs should intern in at least their sophomore and junior summers if not also their freshman summer. Such hands-on experience is invaluable. Even if a student completes a horticultural internship which does not directly support his or her professional aspirations, it is still a learning experience. By planning a second (or third) internship during the summer between junior and senior years, students can gain additional work-related experience prior to graduation. The ideal course of action would be the following. As a university student about to graduate with a degree in ornamental horticulture, you have excelled academically and taken advantage of internship opportunities during the summers of your freshman, sophomore, and junior years. Because of your diligence, good grades, and practical, hands-on work experience, you have built an impressive résumé. You are confident in your abilities. In the fall of your senior year, you started looking into full-time job opportunities having already narrowed down the segment of the Green Industry which would mesh best with your career goals. By Christmas you had held on-campus discussions with recruiters from many businesses in the trade. You also took advantage of on-site interviews with companies that you considered possible opportunities for full-time employment. By early spring you were weighing your options and close to deciding which job offer to accept.

Unfortunately, it is difficult to manage staff if you have had only limited practical hands-on experience. Regardless of which facet of the Green Industry you choose for your career, you need to acquire a broad range of marketable technical skills. Although you may or may not be operating a skid-steer, you will almost certainly be overseeing crew members who do. Do you know the difference between a two- and four-cycle engine? Can you drive a pickup truck with a manual transmission? Are you reasonably familiar with the operation of a chain saw? Have you had experience operating a turf aerator? How familiar are you with the operation of a 48 in. walk-behind mower? Have you driven a large, commercial-size riding mower? Do you know the basic climbing knots used in arboriculture? Do you understand the basics of applying turf herbicides, fungicides, or fertilizers through a rotary spreader? Can you safely mix chemicals and apply them using a backpack or boom sprayer? ❧

Knowing the Industry

Internships give students valuable work experience by introducing them to the day-to-day operation of a horticultural business. Internships help students upgrade their technical skills and become more proficient in them. Another reason for an internship is to gain an understanding of the industry. Most students could spend more time researching their future profession. Let's take the irrigation industry as an example and evaluate its current state and where it is likely to head.

The Green Industry is worth over $65 billion to our nation's economy, and sales have been steadily growing at 10–12 percent for several years. In 2005 sales were expected to grow between only 4 percent and 8 percent. This slightly lower forecast was attributed in part to increased materials cost. The industry is divided between contractors who specialize in commercial work and those who focus on residential customers. Commercial standards have changed over the past 20 years, with more municipalities requiring landscape proposals as part of their building codes. Developers are often required to submit landscape and irrigation plans along with their architectural plans. The residential sector has continued to grow, with new-home construction being particularly healthy in the Sun Belt and western states (Status Report on the Industry, 2005).

Landscape contractors are facing new challenges, as the running of a successful business requires much more than technical proficiency. Companies have broadened the range of products and services they offer. Because most clients prefer to develop a relationship with a single company, many landscape contractors have diversified to offer a more complete range of products and services. Although the transition from operating a specialized business to one offering a full complement of services can be challenging, landscape contractors hope that expanded services will increase customer satisfaction. Companies anticipate increased gross revenues and healthier profit margins into the future (Status Report on the Industry, 2005).

The irrigation industry comprises three major categories. The landscape category includes commercial and residential work and municipalities, schools, and golf courses. The second category is agricultural sales and service; the third is retail. This last category contains garden centers and the do-it-yourself aspect of the trade. Sales of irrigation products for the turf, residentials and commercial market amount to approximately $7 billion annually. Revenue from sale and distribution of irrigation products ranged from 0 to 8 percent by 2005. Diversification of landscape firms has encouraged the growth of larger companies. Until recently, few companies' gross annual income exceeded $5 million. This figure has increased dramatically, as the annual sales of more than a dozen companies in the landscape trade now approach $100 million. Industry sales figures are forecast to increase 5–10 percent in 2006. As a college student, the irrigation industry might be of interest to you as a long-term career choice (Status Report on the Industry, 2005).

Building Your Career

Once hired in a first full-time position, a graduate should continue building his or her career. This means honing managerial skills to aid the climb up the organizational ladder. For example, you can pursue an advanced degree in horticulture, accounting, or business administration or take classes at the local community college to develop your mastery of Spanish. You can take courses offered by your state's Cooperative Extension Service. You can publish articles in popular or professional horticultural magazines or obtain pesticide licensing or commercial driver's license training.

The Importance of Negotiation

One of the primary reasons young supervisors do not advance in an organization is their failure to negotiate successfully. The ability to negotiate is critical to working with peers and also allows young supervisors to be more successful in their work-related responsibilities. Negotiating skills earn respect and appreciation from more senior managers in the organization and are transferable to a wide range of real-life applications. Choose the most opportune time to negotiate. Often the strength of your position depends on timing. Try to have the home-court advantage in any negotiation process. Know the facts and do not be influenced by hearsay or innuendo. Be respectful during any negotiation process and always maintain a positive atmosphere. Be flexible and remember that anger has no place in the negotiation process. Keep your feelings under control and constantly reevaluate each situation. Be honest, be patient, and most of all remain strong and true to your convictions.

Knowing personality traits of others allows prediction of how they will react in a negotiation process. Some individuals strive to please everyone. Others are dominant and controlling. Some try to hide their thoughts;

others express their true feelings. Some are extroverts; others are introverts. Some have high levels of honesty and integrity; others cannot be trusted.

To negotiate successfully, understand what negotiation is and what it is not. First, it obviously takes two or more parties to negotiate. Negotiation is a bargaining process. Think of purchasing a new car. Rarely does one pay full sticker price for a new automobile. You negotiate with the salesperson to get the best possible deal. Although dealers do not expect you to pay full price, they need to make a profit to stay in business. Negotiation is a form of bartering and implies that each party gives up something for the sake of making the deal. Successful negotiation means that each party walks away feeling reasonably comfortable with the final agreement. An equitably negotiated settlement does not have to be a 50–50 trade, but neither should it be 90–10. Too many young supervisors want to win everything and lose nothing in negotiation. Mature managers understand that playing hardball with other supervisors may win the battle, but frequently at the cost of destroying trust with peers. Sound and supportive relationships benefit everyone.

To negotiate effectively,

1. think positively.
2. vary the strategy according to the situation.
3. be clear in prioritizing needs.
4. be flexible.
5. be careful to think through all options before making a commitment.
6. understand the importance of long-term goals.
7. prioritize each important objective.
8. be aware of hidden agendas.
9. be prepared.
10. remain calm.

Suppose you have narrowed your choice for a new pickup truck down to a particular brand and model and know what options are important to you. As you sit down with the salesperson to determine the final price of the vehicle, you need a strategy. What monthly payment can your budget accommodate? The same is true when negotiating a pay raise at the end of the year or negotiating with a trusted employee not to accept a position with a competitor. Establishing parameters is a critical component of any negotiation process. Remember that not every negotiation attempt will end positively. There are times to negotiate and there are times when an individual simply needs his or her space. Remember that stress levels during the height of the spring planting or digging seasons are bound to be high. At this time of year, young supervisors need to be considerate in the way they negotiate and problem solve important issues.

Managing Crisis Situations

Crisis situations often involve some form of negotiation. Effectively handling a true crisis increases your chances of promotion. A *crisis* should not be defined as a small and insignificant issue. Crisis situations are the following: Four weeks before Christmas the temperature-sensing alarm goes off at 2:00 a.m. in your poinsettia greenhouse. It is 20°F outside and poinsettias are very intolerant of cold temperatures. Or you are head superintendent at a golf course hosting a nationally televised tournament next month sponsored by the Ladies Professional Golf Association, and you just found a small but unmistakable infestation of a turfgrass disease such as dollar spot along the back portion of the 18th green. Suppose it is a Friday afternoon in early March and your director of landscape operations informs you she is leaving in 3 weeks. It has been a very mild winter, and all indications are pointing to an early spring season.

To be a successful negotiator,

1. note body language and other telltale signs.
2. fine-tune listening skills.
3. be willing to compromise.
4. be creative in overcoming seemingly unsolvable problems.
5. be courteous and respectful.
6. remain confident.
7. remain positive.
8. treat others with professionalism.

Your busy season is just around the corner, and you had been relying on her to coordinate an array of important responsibilities.

It is not possible to avoid every crisis, so you need to prepare. Crisis situations can more often be avoided when supervisors use sound judgment in the day-to-day management of their operations. If a crisis does develop, take strong and confident action. Be well versed in the circumstances associated with the problem. Keep on top of the situation and do not allow it to grow. Learn and grow from the experience. It is one thing to fail the first time; it is much worse to have the same problem recur.

Understanding Stress

The ability to manage crises and the associated stress is an important component of being a successful supervisor. In managing work-related stress, be aware that stress has different effects. Pupils dilate, heart rate increases, bodies increase adrenaline production, and sweat glands kick into high gear. Jobs that have high levels of stress can increase employee absenteeism and decrease productivity. Stressful work environments impair how staff work and communicate. A wholesale B&B nursery struggling to meet spring digging orders is a stressful environment. A large landscape design and installation contractor with a looming deadline also creates stress. Managers need to be proactive in managing stress for themselves and staff. The different facets of the horticulture industry have their own peak production cycles. Bringing a poinsettia crop to market occurs at a much different time of year than bluegrass turf production at a commercial sod farm. The May prom season is a busy time at any florist shop; Halloween is for a pumpkin grower. Try to guard against reacting too emotionally when problems do arise, and do not allow stressful situations to negatively influence decision making. Remain confident in your job-related abilities and maintain production levels at acceptable levels. Work-related stresses increase as you climb the managerial ladder and responsibilities increase. Leaving the employ of an established company and starting your own business can be very stressful as well.

Managing Stress

Work-related stress increases as you grow in your career and will be compounded by stress in your personal life as you establish your own family.

Watch for signs of stress in staff. Identify those who seem to be more affected by work-related responsibilities and become easily stressed. Determine whether your company offers stress-management seminars to employees. Monitor employee absenteeism and job-related accidents to determine if they are stress related. They might be if changes significantly alter employee routines or are instituted with little forewarning. Often, job-related stress can be significantly lessened through sound and proactive management.

Two important practices to initiate early in your professional career include eating a balanced diet and exercising regularly. Personal habits formed early in life reap huge dividends as a career matures into middle and upper managerial positions.

Managing Stress Through Organization

Become adept at organizing the work space you occupy. Clean out your paper files and e-mails much more often than on a semiannual basis. Organizing your work space might take time, but your ultimate goal is to be as efficient as possible. Effective managers cleanse, clear, and throw away. A work environment in the Green Industry may spread well beyond an office. If you are the assistant production manager at a B&B wholesale nursery, what does the seat of your pickup truck look like? It doesn't need to be swept and clean, but your effectiveness is certainly being compromised if you can't see the floor for the litter of work lists, client telephone messages, digging orders, and invoices. Disorganization creates inefficiency that puts unnecessary stress on you and your staff.

Unnecessary worry encourages stress. Rationally think through problematic situations. Consider viable alternatives and arrive at an adequate solution. By doing so, you will feel more self-assured and in control of the issue. Do not avoid confronting important problems head-on. Avoiding uncomfortable situations only exacerbates anxiety. Confronting situations encourages the recent graduate to feel that he or she has adequate control over work-related problems or issues. Worry-prone supervisors should not feel helpless if they do not have immediate answers to every problem that comes their way. Many problems demand sufficient time to arrive at an equitable solution.

Managing stress in the workplace includes

1. emulating the techniques of those who manage stress effectively.
2. delegating.
3. balancing personal and professional life.
4. rewarding yourself with positive reinforcements after reaching significant goals.
5. working *smart* as well as working *hard*.
6. not being afraid of change.
7. being realistic about your personal and professional aspirations.

CONCLUSION

Rather than providing guidelines in this chapter for students to use in achieving career goals, I chose a more realistic approach. Young managers usually become proficient in their time-management, decision-making, delegation, and leadership skills. They master the ability to plan and motivate their staff. It is deficient negotiation skills, inability to deal with crises and stress, and disorganization that often keeps them from being promoted.

QUESTIONS FOR DISCUSSION

1. The ideal horticulture senior presented at the chapter's beginning had started the interview process by Christmas and by early spring was considering full-time-employment job offers. Do you think that this is a realistic model for a top-college senior to follow?

To manage stress,

1. exercise regularly and maintain a healthy diet.
2. develop a support system both for self and staff.
3. keep a close eye on staff absenteeism.
4. schedule adequate time to complete jobs and tasks.
5. set realistic goals and objectives.
6. remain positive.
7. be proactive in dealing with problems.
8. be able to say no.
9. be realistic in your capabilities to follow through.

2. What benefit might there be to starting to forge a career in the Green Industry early in one's college career?
3. Do you agree that a quality internship experience is an important component of planning a career within the horticulture profession?
4. Do multiple college internships make a student much more prepared to meet his or her long-term professional goals?
5. How important are internships in helping students decide which facet of the horticulture profession to pursue?
6. Many if not most of the full-time positions that graduating horticulture seniors accept are offered because the individual had interned previously with the firm. How does that affect you?
7. Is being skilled in negotiation important to a young supervisor's promotability within the horticulture profession?
8. Why make some concessions when negotiating?
9. Why should each side feel reasonably comfortable with the compromises in a negotiated settlement?
10. Why should a young supervisor remain flexible when negotiating with other managers within an organization?
11. List the suggestions presented in this chapter which would best assist you in working through crisis situations.
    What can you add?
12. Why is stress management necessary for a forward-looking and proactive supervisor?
13. Can the effectiveness by which a young supervisor manages stressful situations significantly impact his or her upward climb in an organization?
14. How does an organized work environment assist young supervisors in being more efficient in their day-to-day activities?

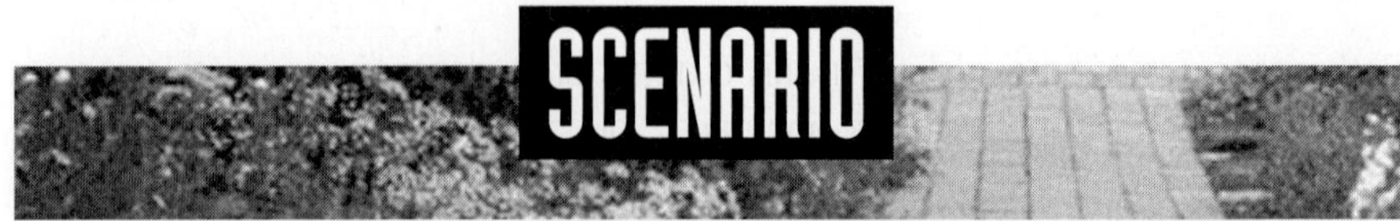

Carl Richard's Full-time Employment Experience at Harbor View Country Club

Carl Richards was hired as a full-time member of the grounds staff at Harbor View Country Club after graduation from a well-known university in the southeast. Harbor View was located just outside Myrtle Beach, South Carolina. Carl had interned at Harbor View during the summer between his junior and senior years in college and was offered a full-time position with the club due to his above-average work ethic and long-term interest in the golf course industry. The club had been chosen to host the state's Senior Women's Amateur Championship. Thrilled at the opportunity, Harbor View management was especially pleased that their new 18-hole championship course would be featured for the event. Although the tournament was not scheduled until the following June, the grounds staff was well aware of its increased responsibilities in fine-tuning the course.

Carl had been a full-time employee at Harbor View for just over a year and reported directly to Ms. Sandra Roberts. Ms. Roberts was an exceptionally talented assistant superintendent who had been with Harbor View for nine full seasons. In preparation for the tournament, Ms. Roberts was to oversee a wide array of responsibilities that would improve the course's aesthetics and make it more challenging for the golfers. Improvements ranged from recontouring sand traps and bunkers to major annual and perennial plantings surrounding the clubhouse and finishing holes of the course. Meadow plantings of summer-blooming wildflowers were also planned to accentuate the broad sweeps of land adjoining the 9th, 14th, 17th, and 18th fairways.

Carl was a diligent and motivated employee who had a sound grasp of turf-related principles and techniques. Carl had a tendency to be high strung, especially when involved in a crisis. As soon as Harbor View received the news that it would be hosting the tournament, Ms. Roberts began planning for the landscape renovation projects she would oversee. As an experienced supervisor, she knew the championship would arrive quickly and she needed to make progress on the many renovations required for the tournament.

Over the next few months, Ms. Roberts and her crew recontoured sand traps and bunkers and planted wildflowers along the 17th and 18th fairways. Rather than continue planting the sweeps of wildflowers along the 9th and 14th fairways, Ms. Roberts decided to start renovating the landscapes adjacent the clubhouse and surrounding areas. This allowed more efficient sharing of trucks, tractors, and other equipment with Trang Lam, the second assistant superintendent at the club. As Ms. Roberts' crews began renovating the clubhouse landscape, she assigned the planting of a large array of summer-blooming herbaceous perennials to Carl and two college interns under his supervision. After receiving clear and specific instructions, Carl had the interns sign out the necessary digging tools from the shop and meet him at the planting site. In the meantime, he went to secure a truck to pick up the first load of perennials for the clubhouse project. As Ms. Roberts returned to her office, she noticed Carl exiting the fuel area in one of the department's trucks and traveling toward the clubhouse. Although he drove off at a safe speed, the anxious look on his face concerned her. She was soon met by assistant superintendent Lam, who had just seen Kenyatta Mason, one of his workers and a full-time member of the grounds staff, near the gas pumps in an obviously emotional discussion with Carl. Trang was concerned that Carl drove off in the last remaining truck without transporting tools and a tarp for an important task assigned to Kenyatta, who had no choice but to use one of the remaining gas-powered cargo carts.

As Ms. Roberts left her office in a gas-powered cart, she noticed Kenyatta in the distance planting perennials with another intern near the 14th tee. Ms. Roberts stopped by to discuss Kenyatta's recent encounter with Carl. Kenyatta described how she had requested Carl to drop off some hand tools and a tarp at a location out of sight of patrons playing the course. Doing so would have saved her and the intern considerable time. Kenyatta was a level-headed and competent employee and her request was reasonable.

As Ms. Roberts approached Carl, she could see he was not calm. After looking over the perennial plantings that Carl and his intern had completed, she asked Carl to sit down with her in the cart and explain his confrontation with Kenyatta. Carl explained that he did not want to take time away from his responsibilities to drop off the tools and tarp for Kenyatta. He felt pressured to complete the plantings near the clubhouse even though the championship was not imminent. He admitted getting emotionally caught up in working toward the completion of such a large and important landscape project.

Although Carl had not always dealt well with stress-related situations, he had never before become so agitated that he argued with a fellow staff member. Ms. Roberts explained to him that his outburst was not acceptable. She told Carl to think through the consequences of his actions and asked what disciplinary measures might be in order. Wanting Carl to understand the seriousness of his actions, she asked him to imagine that his outburst had occurred within earshot of a club member. Ms. Roberts was trying to have Carl consider the importance of an employee's appropriate conduct. After a moment of thought, Carl said that a written apology to Kenyatta would certainly be in order and promised that he would never repeat his unacceptable action. After discussing the situation with Harbor View's director of human resources, Ms. Roberts decided to have Carl attend a stress-management seminar at the local community college. She hoped that such an approach would assist him in dealing with the day-to-day responsibilities of his position. Harbor View would cover the cost of the seminar, which met weekly for 8 weeks. After the seminar, Carl was better able to prioritize his tasks and keep the overall responsibilities of his position more firmly in focus. He was better able to conduct himself in a calm, professional manner when involved in a crisis.

QUESTIONS FOR DISCUSSION

1. In view of Carl's high-strung personality, how would you have managed Carl on a day-to-day basis?
2. As Carl's supervisor, would you select tasks that best fit Carl's personality, or would you simply manage him the same as the rest of your staff?
3. Should Carl's attributes as a capable and motivated employee outweigh his limited ability to deal with stressful situations?

4. Do you agree with Sandra Roberts's handling of the situation between Carl Richards and Kenyatta Mason?
5. If you had been Carl's supervisor, how would you have managed the confrontation between him and Kenyatta?
6. Do you see Carl eventually being promoted to a crew leader's at Harbor View?
7. Do you think Carl could rise to an assistant superintendent's position at the club?
8. Will Carl be able to survive long-term as a full-time employee at Harbor View?
9. Overall, Carl Richards was a diligent and dedicated full-time employee at Harbor View. Do you agree with Sandra Roberts's decision to have Carl attend a stress-management seminar at the local community college?
10. How difficult is it to retain diligent and dependable full-time staff at an upscale golf club such as Harbor View?
11. What is Carl's overall value as a full-time employee at Harbor View?

CRITIQUE

Carl Richards had difficulty in dealing with stress. Carl's obvious inability to work through stressful situations had clouded his judgment and he was unprofessional with his fellow horticultural employee at the club.

Low-level supervisors seldom have to deal with many highly stressful situations. However, a stressful situation for one employee may hold little or no concern for another. Keeping an eye out for emotional stresses within a workforce is important to the long-term management of many types of personalities. Carl did not manage his emotions well when Kenyatta asked him to assist her in transporting the tools and tarp and was wrong to lose his temper over the situation. Carl's outburst was most likely the result of pent-up frustrations. Ms. Roberts was correct to discuss the situation with Kenyatta Mason. Kenyatta was an honest and upright employee who had no ax to grind with Carl. She had made a reasonable request of Carl, and he had reacted unacceptably. Remember that Kenyatta's sole transportation choice was a relatively slow, gas-powered cargo cart. Supervisors should look closely at employees who become stressed over relatively insignificant situations. Sometimes an emotional outburst over something trivial indicates stress caused by something of more consequence. Ms. Roberts encouraged Carl to think through the ramifications of his actions rather than harshly chastising him. Ms. Roberts wanted Carl to see a bigger picture in association with his actions. Carl was a well meaning and dependable employee, but he did not have the technical or managerial capability to become an assistant superintendent at an exclusive course like Harbor View. Carl probably had the potential of developing into a crew leader at Harbor View, assuming he could overcome his stress-related anxieties.

Assistant superintendent Roberts's decision to seek the guidance of Harbor View's director of human resources was a wise and managerially mature decision. The 8 week stress-management seminar was also an excellent alternative to the more common type of disciplinary action that Carl could also have faced.

BIBLIOGRAPHY

2006 Wage & Benefits Survey [A Staff Report]. (2007, January 15). *American Nurseryman*, 205(2), 33–44.

Adams, J. L. (1986). *The care and feeding of ideas: A guide to encouraging creativity*. Reading, MA: Addison-Wesley Publishing Company.

Belasco, J. A., & Stayer, R. C. (1993). *Flight of the buffalo: Soaring to excellence, learning to let employees lead* (p. 19). New York: Warner Books.

Cairo, J. (1995). *Motivation and goalsetting: The keys to achieving success* (pp. 31–32). Shawnee Mission, KS: National Press.

Covey, S. R. (1990). *The 7 habits of highly successful people: Powerful lessons in personal change* (pp. 149, 171). New York: Fireside.

Davidson, H., Peterson, C., & Mecklenburg, R. (1994). *Nursery management: Administration and culture* (pp. 159, 476). Englewood Cliffs, NJ: Prentice Hall Career & Technology.

Ehringer, A. G. (1995). *Entrepreneurs talk about decision making* (pp. 2, 54–55). Santa Monica: Merritt Publishing, A Division of The Merritt Company.

Goleman, D., Boyatzis, R., & McKee, A. (2002). *Primal leadership: Realizing the power of emotional intelligence* (p. 15). Boston: Harvard Business School Press.

Grossman, J., & Parkinson, J. R. (2002). *Becoming a successful manager: How to make a smooth transition from managing yourself to managing others* (pp. 147–157). New York: McGraw-Hill.

Harvard Business Essentials (HBS). (2004). *Manager's toolkit: The 13 skills manager's need to succeed* (pp. 218, 238–240, 257–258). Boston: Harvard Business School Press.

Hrebiniak, L. (2005). *Making strategy work: Leading effective execution and change* (pp. 228–229). Upper Saddle River, NJ: Wharton School Publishing.

Kotter, J. (1999). *What leaders really do* (pp. 51–53, 91). Boston: Harvard Business School Press.

Marcus, L. (2007, February 15). Employee motivation. *American Nurseryman*, 205(4), 32–35.

Milo, F. (1989). *How to run a successful meeting in half the time* (pp. 82–83). New York: Simon and Schuster.

Mulhern, B. (2004, December). Getting the message across: Communicating safety to employees with limited English skills. *Nursery Management & Production*, 20(12), 59–62.

National Institute of Occupational Safety and Health. (2005, January). Safe lifting and carrying techniques. *Turf North*, 18(1), C6.

Reichheld, F. (1996). *The loyalty effect: The hidden force behind growth, profits and lasting value* (p. 33). Boston: Harvard Business School Press.

Rogers, C., & Roethlisberger, F. J. (Eds.) (1952/1991). Barriers and gateways to communication. In *Business classics: Fifteen key concepts for managerial success* (p. 45). Boston: Harvard Business School Press.

Status Report on the Industry [Staff Report]. (2005, November). *Irrigation & Green Industry*, pp. SR4–SR15.

Townsend, P., & Gebhardt, J. E. (1992). *Quality in action* (pp. 25–26). New York: John Wiley & Sons.

Velasquez, M. (2006, March). The first and second Hispanic-Latino wave. *Alna Today*, 18(2), 9, 15.

Wayland, R. E., & Cole, P. M. (1997). *Customer connections: New strategies for growth* (p. 103). Boston: Harvard Business School Press.

Whiteley, R., & Hessan, D. (1996). *Customer-centered growth: Five proven strategies for building cooperative advantage* (pp. 10–17). Reading, MA: Addison-Wesley.

Willingham, R. (1992). *Hey I'm the customer* (p. 9). New York: Prentice Hall.

Wilson, D. (2001). Performance appraisal. In M. J. Fleming & J. B. Wilson (Eds.), *Effective HR measurement techniques* (pp. 77–96). Alexandria, VA: Society for Human Resource Management.

Wisniewski, N. (2005, October). 2005 State of the industry report: Working smarter. *Lawn & Landscape*, 26(10), S1–S23.

INDEX

A

Accountability, 75
Authoritative leadership, 71
Average annual revenue, 85
Average net profit, 85

B

Balance sheet, 88–89
Behavior
  innate behavior, understanding, 29
  negative employee, 151
Body language, 51
Brevity, 49
Budgeting process, 86–88
  incremental budget, 87
  rolling budget, 87
  zero-based budget, 87
Business plan, 90

C

Candidate selection, for interview, 138
Career path, 156–166
  building, 158
  crisis situations, managing, 159–160
  knowledge of industry, 157–158
  negotiation, importance of, 158–159
  stress
    managing, 160–161
    understanding, 160
Cash flow
  positive, 128
  statement, 89–90
Change, 127–135
  financial costs, of change, 127
  initiating, 130
  managing, 129–130
    difficulty in, 128–129
  organizational change, managing, 128
  trust, building, 130–131
Clarity, 49
Communication, 47, 72, 129, 131, 147, 151
  barriers to, 51–52
  office, 17
  oral, 49–50
  telephone, 50
  written, 49
Compensation, 150
  package, 137, 138
Competitiveness, 6
Coordination, 73–75
Crisis situations, managing, 159–160
Customer, 38
Customer base, 86
Customer service
  and fiscal health, 85–86

D

Decision making
  benefits of mentor, 39
  and future, 38–39
  informed decision making, 39–40
  and long-term goals, 37–38
  risk-taking consequences, 40–41
  tough decisions, making, 41
Delegating authority, 71
  accountability, 75
  coordination, 73–75
  managerial support, 75
  productivity, 73–75
Delegation, 17, 71–72
  and communication, 72
  efficient, 73
  and leadership, 71–72
Delegative leadership, 72
Difficult employees, motivating, 29–30
Dressing, 118–126
  first impressions, 119–120
  and organizational policy, 118–119

E

Education
  to hone official skills, 6
Effective meetings. *See* Meetings
Employee evaluation. *See* Staff performance, evaluating
Employee motivation, 26
Employee understanding, by supervisor, 28–30

F

Financial statements
  balance sheet, 88–89
  cash flow statement, 89–90
  income statement, 89
Fiscal health, managing, 84
  budgeting process, 86–88
  customer service, 85–86
  financial statements, 88–90
  landscape industry finances, 85

G

Goal setting, 2–3, 16–17, 109–110
  problem solving, 3
Green industry, motivation in, 30–31

H

Hiring. *See* Staff, hiring
Honesty, 28, 47–49

I

Income statement, 89
Incremental budget, 87
Inefficient meetings, 100
Informed decision making, 39–40
Innate behavior, understanding, 29
Integrity, 28, 47–49
Internships, 157
Interview, 138–139
Irrigation industry, 158

J

Job satisfaction, 147

L

Landscape contractors, 157, 158
Landscape industry finances, 85
Leadership, 58, 71–72
  authoritative, 71
  delegative 72
  participatory, 72
  true, 58–63
Long-term goals
  and decision making, 37–38
Long-term planning, 111

M

Managerial support, 75
Meetings, 97–107
  concluding, 102
  managing, 101–102
  planning of, 99–101
  productivity of, 99
  professionalism in, 101
Mentoring, 4–5, 61
  benefits, 39

Motivation, 25
  of difficult employees, 29–30
  employee understanding, by supervisor, 28–30
  in green industry, 30–31
  innate behavior, understanding, 29
  personalization of, 30
  plan, developing, 27–28
  of staff, 26–27

N

Negative employee behavior, 151
Negotiation, 158–159
Net revenue, 85

O

Objectivity, 38
Office communication, 17
Oral communication, 49–50
Organization, transforming, 59
Organizational change, managing, 128
Own business, 90

P

Participatory leadership, 72
Performance appraisals, fine-tuning, 150–151
Personality traits, knowing, 158–159
Personalizing motivation techniques, 30
Persuasion, 50, 63–64
Planning, 13–16, 108–117
  components, 108–109
  developing, 27–28
  goal setting, 109–110
  of meeting, 99–101
  time management, 110–111
  to-do-list, 110
Problem solving, 75
  goal setting, 3
  and staff management, 2–3
Procrastination, 17–19, 110
Productivity, 73–75
Professionalism, 101

R

Rolling budget, 87

S

Scheduling, 138
Selectivity, 49
Staff, hiring, 136–145
  advertisement of position, 137–138
  candidate selection, for interview, 138–139
  hiring strategy, developing, 137
  needs, defining, 136–137
Staff management, 2
  effectiveness of, 4
  motivation, 26–27
  problem solving and, 2–3
Staff performance, evaluating, 146–155
  compensation, 150
  elements, 147–150
  job satisfaction, 147
  negative employee behavior, 151
  performance appraisals, fine-tuning, 150–151
Stress
  managing, 160–161
  understanding, 160
Supervisors, 2, 3, 6, 13–15, 18, 25–26, 158. *See also* Change
  decision-making, ability of, 39–41
  delegating authority, 72, 73, 74
  employees, understanding, 28–30
  leadership, 61, 62
  motivating, 28
    quality, 27
  and staff, relationship, 50
  trustworthiness, 48
  working with, 5
  written communication skill, 49

T

Tasks, prioritizing, 12–13
Teamwork, 3–4
Telephone communication, 50
Time management, 11–24, 110–111
  core principles, 13
  goal setting, 16–17
  organization, 19
  planning, 13–16
  tasks, prioritizing, 12–13
To-do-list, 110
Total average revenue, 85
Tough decisions, making, 41
True leadership, 58–59
  characteristics, 61–63
  principles, 60–61
Trust, 28, 47–49, 72
  building, 130–131

W

Written communication, 49

Z

Zero-based budget, 87